ADVANCED HOME WIRING

Updated 6th Edition

Current with 2023–2026 Electrical Codes

COOL SPRINGS PRESS

Quarto.com

© 1992–2024 Quarto Publishing Group USA Inc.

Fifth edition published in 2018; Fourth edition 2015; Third edition 2012; Second edition 2008

First Published in 1992 by Creative Publishing international, now Cool Springs Press, an imprint of The Quarto Group, 100 Cummings Center, Suite 265-D, Beverly, MA 01915, USA. T (978) 282-9590 F (978) 283-2742

Cool Springs Press titles are also available at discount for retail, wholesale, promotional, and bulk purchase. For details, contact the Special Sales Manager by email at specialsales@quarto.com or by mail at The Quarto Group, 100 Cummings Center, Suite 265-D, Beverly, MA 01915, USA.

28 27 26 25 24   1 2 3 4 5

ISBN: 978-0-7603-8818-1

Digital edition published in 2024
eISBN: 978-0-7603-8819-8

Acquiring Editor: Mark Johanson
Project Manager: Jordan Wiklund
Art Director: Brad Springer
Cover Designer: Brad Springer
Layout: Danielle Smith-Boldt
Technical Reviewer: Bruce Barker
New Photography: Bruce Barker
New Illustrations: Ada Keesler on pages
    52 (upper right and lower right), 53–57

Printed in China

NOTICE TO READERS

For safety, use caution, care, and good judgment when following the procedures described in this book. The publisher and BLACK+DECKER cannot assume responsibility for any damage to property or injury to persons as a result of misuse of the information provided.

The techniques shown in this book are general techniques for various applications. In some instances, additional techniques not shown in this book may be required. Always follow manufacturers' instructions included with products, since deviating from the directions may void warranties. The projects in this book vary widely as to skill levels required: some may not be appropriate for all do-it-yourselfers, and some may require professional help.

Consult your local building department for information on building permits, codes, and other laws as they apply to your project.

Contents

Advanced **Home Wiring**

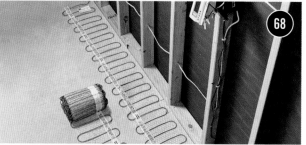

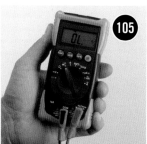

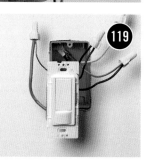

Introduction

Experienced home electricians understand the need to keep up with changes in the world of wiring. Wiring projects, and more advanced projects in particular, almost always require a permit from your municipality and typically an on-site inspection or two as well. If you aren't up-to-date with wiring codes, there is a likelihood that your project will not pass inspection. But beyond the practicality of passing inspections, the codes that govern wiring practices are updated for good reason: they improve safety. And when you're talking about your own home and family, that's worth paying attention to.

This newest edition of BLACK+DECKER *Advanced Home Wiring* has been reviewed and revised to reflect the changes to wiring code published in the 2023 edition of the National Electrical Code (NEC), which is updated every three years.

Most of the advanced wiring projects featured in this book involve the installation of new circuitry, panel upgrades, or troubleshooting with diagnostic equipment. Among the high-level projects: upgrading the grounding and bonding on your new 200-amp or larger home circuit; installing an automatic transfer switch for your backup power supply; wiring a room addition; using a multimeter to precisely locate an open neutral in a home circuit; and installing new

"smart" thermometers that are programmable and/or linked to the internet.

Because the projects found in this book are advanced in nature, do not attempt any of them unless you are confident in your abilities. Consult a licensed professional electrician if you have any concerns—in many cases your best solution might be to do some of the work yourself, such as pulling new sheathed cable through walls, and to have the electrician do the other work, such as making the connections. But keep in mind that home wiring can be a fun and fascinating pursuit, and successfully accomplishing a major project is personally gratifying and can also save you substantial amounts of money.

Wiring Safety

Safety should be the primary concern of anyone working with electricity. Although most household electrical repairs are simple and straightforward, always use caution and good judgment when working with electrical wiring or devices. Common sense can prevent accidents.

The basic rule of electrical safety is: Always turn off power to the area or device you are working on. At the main electrical panel or subpanel, remove the fuse or shut off the circuit breaker that controls the circuit you are servicing. Then check to make sure the power is off by testing for power with a voltage tester.

TIP: Test a live circuit with the voltage tester to verify that it is working before you rely on it.

Restore power only when the repair or replacement project is complete.

Follow the safety tips shown on these pages. Never attempt an electrical project beyond your skill or confidence level.

Shut power OFF at the panel or subpanel where the circuit originates before beginning any work.

Create a circuit index and affix it to the inside of the door of all panels and subpanels. Update it as needed.

Confirm power is OFF by testing at the outlet, switch, or fixture with a voltage tester.

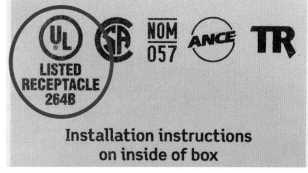

Use only approved electrical parts or devices. These devices have been tested for safety by Underwriters Laboratories (UL) and similar testing agencies.

Wear rubber-soled shoes while working on electrical projects. On damp floors, stand on a rubber mat or dry wooden boards.

Use fiberglass or wood ladders when making routine household repairs near the service mast.

Extension cords are for temporary use only. Cords must be rated for the intended usage.

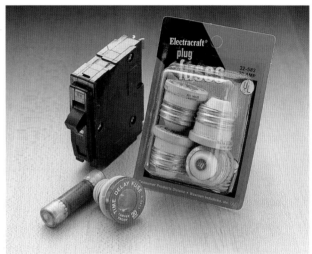

Breakers and fuses must be compatible with the panel manufacturer and match the circuit capacity.

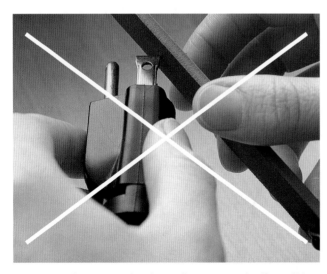

Never alter the prongs of a plug to fit a receptacle. If possible, install a new grounded receptacle.

Do not penetrate walls or ceilings without first shutting off electrical power to the circuits that may be hidden.

Planning Your Project

Careful planning of a wiring project ensures you will have plenty of power for present and future needs. Whether you are adding circuits in a room addition, wiring a remodeled kitchen, or adding an outdoor circuit, consider all possible ways the space might be used, and plan for enough electrical service to meet peak needs.

For example, when wiring a room addition, remember that the way a room is used can change. In a room used as a spare bedroom, a single 15-amp circuit provides plenty of power, but if you ever choose to convert the same room to a family recreation space, you will need additional circuits.

When wiring a remodeled kitchen, it is a good idea to install circuits for an electric oven and countertop range, even if you do not have these electric appliances. Installing these circuits now makes it easy to convert from gas to electric appliances at a later date.

A large wiring project adds a considerable load to your main electrical service. In some homes, an upgrade of the electrical service or the main service panel is needed before new wiring can be installed. For example, some homeowners will need to replace an older 60-amp electrical service with a new service rated for 100 amps or more. This is a job for a licensed electrician but is well worth the investment. In other cases, the existing main service provides adequate power, but the main circuit breaker panel is too full to hold any new circuit breakers. In this case it is necessary to install a circuit breaker subpanel to provide room for hooking up added circuits. Installing a subpanel is a job most homeowners can do themselves.

This chapter gives an easy five-step method for determining your electrical needs and planning new circuits.

Five Steps for Planning a Wiring Project

Examine your main service panel. The amp rating of the electrical service and the size of the circuit breaker panel will help you determine if a service upgrade is needed. If there is a question about a service upgrade, an electrician should perform a load calculation.

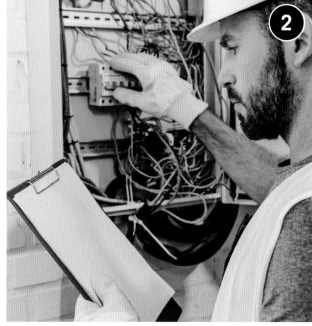

Learn about codes. The National Electrical Code (NEC), and local electrical codes and building codes, provide guidelines for determining how much power and how many circuits your home needs. Your local electrical inspector can tell you which regulations apply to your job.

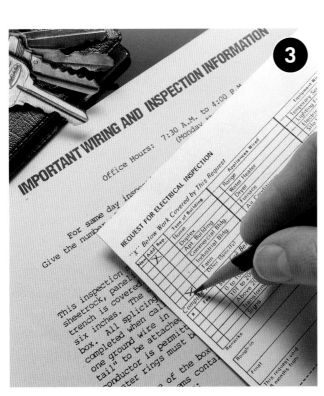

Prepare for inspections. Remember that your work must be reviewed by your local electrical inspector. When planning your wiring project, always follow the inspector's guidelines for quality workmanship.

Evaluate electrical loads. New circuits put an added load on your electrical service. Make sure that the total load of the existing wiring and the planned new circuits does not exceed the service capacity or the capacity of the panel.

Draw a wiring diagram and get a permit. This wiring plan will help you organize your work.

Examine Your Main Service Panel

The first step in planning a new wiring project is to look in your main electrical panel and find the size of the service by reading the amperage rating on the main circuit breaker. As you plan new circuits and evaluate electrical loads, knowing the size of the main service helps you determine if you need a service upgrade.

Also look for open circuit breaker slots in the panel. The number of open slots will determine if you need to add a circuit breaker subpanel.

Find the service size by opening the main service panel and reading the amp rating printed on the main circuit breaker. This method works when there is one main circuit breaker or fuse block. Some houses have multiple services disconnects. In these cases, contact an electrician to determine your service size. In most cases, 100-amp service provides enough power to handle the added loads of projects such as the ones shown in this book. A service rated for 60 amps or less should be upgraded.

NOTE: In some homes the main circuit breaker is located in a separate box. The main circuit breaker may be located outside in newer homes.

Older panels use fuses instead of circuit breakers. Have an electrician replace this type of panel with a circuit breaker panel that provides enough power and enough open breaker slots for the new circuits you are planning.

Look for open circuit breaker slots in the main circuit breaker panel or in a circuit breaker subpanel, if your home already has one. You will need one open slot for each 120-volt circuit you plan to install and two slots for each 240-volt circuit. If your main circuit breaker panel has no open breaker slots, install a subpanel to provide room for connecting new circuits. Do not install more circuit breakers in any panel than are allowed by the panel manufacturer. The maximum number of circuit breakers is usually encoded in the panel model number. Tandem (half-height) circuit breakers count as two circuit breakers.

Learn About Codes

To ensure public safety, your community requires that you get a permit to install new wiring and have the work reviewed by an inspector. Electrical inspectors use the National Electrical Code (NEC) as the primary authority for evaluating wiring, but they also follow the local building code and electrical code standards.

Most communities use a version of the NEC that is not the most current version. Also, many communities make amendments to the NEC, and these amendments may affect your work.

As you begin planning new circuits, call or visit your local electrical inspector and discuss the pro ject with him or her. The inspector can tell you which of the national and local code requirements apply to your job and may give you a packet of information summarizing these regulations. Later, when you apply to the inspector for a permit, he

or she will expect you to understand the local guidelines as well as a few basic NEC requirements.

The NEC is a set of standards that provides minimum safety requirements for wiring installations. It is revised every three years. The national code requirements for the projects shown in this book are thoroughly explained on the following pages. For more information, you can find copies of the current NEC, as well as a number of excellent handbooks based on the NEC, at libraries and bookstores. You can also access a free (read-only) copy of the NEC at NFPA.org.

In addition to being the final authority of code requirements, inspectors are electrical professionals with years of experience. Although they have busy schedules, most inspectors are happy to answer questions and help you design well-planned circuits.

Basic Electrical Code Requirements

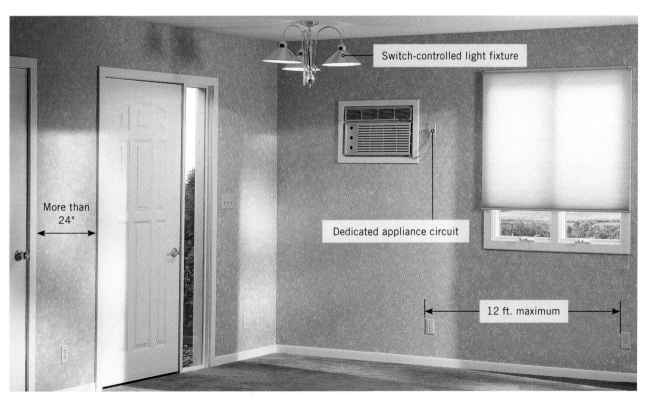

Electrical code requirements for living areas: Living areas need at least one 15-amp or 20-amp basic lighting/receptacle circuit for each 600 sq. ft. of living space and should have a dedicated circuit for each type of permanent appliance, such as an air conditioner, or a group of baseboard heaters. Receptacles on basic lighting/receptacle circuits should be spaced no more than 12 ft. apart. Many electricians and electrical inspectors recommend even closer spacing. Any wall more than 24" wide also needs a receptacle. Every room should have a wall switch at the point of entry to control either a ceiling or wall-mounted light or plug-in lamp. Kitchens and bathrooms must have a ceiling or wall-mounted light fixture.

BY MATERIAL
Panels
- Maintain a minimum 30" wide by 36" deep of clearance in front of the panel.
- Match the amperage rating of the circuit when replacing circuit breakers and fuses.
- Use handle ties on all 240-volt breakers and on 120-volt breakers protecting multi-wire branch circuits.
- Close all unused panel openings.
- Label each fuse and breaker clearly on the panel.

Electrical Boxes
- Use boxes that are large enough to accommodate the number of wires and devices in the box.
- Install all junction boxes so they remain accessible.
- Leave no gaps greater than ⅛" between wallboard and the front of electrical boxes.
- Place receptacle boxes flush with combustible surfaces.
- Leave a minimum of 3" of usable cable or wire extending past the front of the electrical box.

Wires & Cables
- Use wires that are large enough for the amperage rating of the circuit.
- Drill holes at least 2" from the edges of joists. Do not attach cables to the bottom edge of joists.
- Do not run cables diagonally between framing members.
- Use nail plates to protect cable that is run through holes drilled or cut into studs less than 1¼" from the front edge of a stud.
- Do not crimp cables sharply.
- Contain spliced wires or connections entirely in a plastic or metal electrical box.
- Use wire connectors to join wires.
- Secure cables within 8" of an electrical box and every 54" along its run.
- Leave a minimum ¼" (maximum 1") of sheathing where cables enter an electrical box.
- Clamp cables and wires to electrical boxes with approved clamps. No clamp is necessary for one-gang plastic boxes if cables are secured within 8".
- Connect only a single wire to a single screw terminal. Use pigtails to join more than one wire to a screw terminal.

Switches
- Use a switch-controlled receptacle in rooms without a built-in light fixture operated by a wall switch.
- Use three-way switches at the top and bottom on stairways with six risers or more.
- Use switches with grounding screws with plastic electrical boxes.
- Locate all wall switches within easy reach of the room entrance, ideally not behind the door.
- Install a neutral wire in switch boxes.
- Use black or red wires to supply power to switched devices.

Receptacles
- Install receptacles on all walls 24" wide or greater.
- Install receptacles so a 6-foot cord can be plugged in from any point along a wall or every 12 ft. along a wall.
- Use three-slot, grounded receptacles for all 15- or 20-amp, 120-volt branch circuits.
- Include a switch-controlled receptacle in rooms without a built-in light fixture operated by a wall switch.
- Install GFCI-protected circuits in bathrooms, kitchens, laundry rooms, garages, crawl spaces, unfinished basements, and outdoor receptacle locations.
- Install one 15-amp or 20-amp, 120-volt, GFCI-protected, receptacle for each parking space in a garage. Use the garage receptacle circuit only for receptacles in the garage and for receptacles located on garage exterior walls.
- Install at least one 120-volt, GFCI-protected, receptacle in each unfinished basement area.
- Install a 120-volt receptacle within 25 feet from HVAC equipment such as furnaces, boilers, and condensers.

Light Fixtures
- Use mounting straps that are anchored to the electrical boxes to mount ceiling fixtures.
- Keep non-IC-rated recessed light fixtures 3" from insulation and ½" from combustibles.
- Include at least one switch-operated lighting outlet in every habitable room, kitchen, bathroom, basement, hallway, stairway, attached garage, and attic and crawl space area that is used for storage or that contains equipment that requires service. This outlet may be a switched receptacle in areas other than kitchens and bathrooms.
- Do not install dimmer switches on interior stair lights unless a dimmer is installed on all switches controlling these lights.

AFCI and GFCI Protection
- Extending a branch circuit or adding a new branch to install new receptacles, lights, switches, or equipment requires a permit. The electrical inspector may require that you install arc-fault protection on the entire circuit and may require that you install GFCI protection where

currently required. GFCI protection may be required on 120-volt and 240-volt circuits. Check with the electrical inspector before starting such projects.

Grounding
- Ground receptacles by connecting receptacle grounding screws to the circuit grounding wires.
- Use switches with grounding screws whenever possible. Always ground switches installed in plastic electrical boxes and all switches in kitchens, bathrooms, and basements.

BY ROOM
Kitchens/Dining Rooms
- Install at least two 20-amp small appliance receptacle circuits.
- Install dedicated 15-amp, 120-volt circuits for dishwashers and food disposals (required by many local codes). The dishwasher circuit should be GFCI protected.
- Install GFCI protection for all countertop receptacles; and for those within 6 feet from the sink. The 6 feet from sink rule includes the refrigerator receptacle, unused receptacles under the sink, and receptacles along walls such as below a breakfast bar.
- Position receptacles for appliances that will be installed within cabinets, such as microwaves or food disposals, according to the manufacturer's instructions.
- Install receptacles at all countertops and work surfaces wider than 12".
- Space receptacles a maximum of 48" apart above countertops and closer together in areas where many appliances will be used.
- Locate receptacles on the wall above the countertop not more than 20" above the countertop.
- Install at least one receptacle not more than 12" below the countertop to serve the first 9 sq. ft. of islands and peninsulas. Install at least one receptacle to serve each additional 18 sq. ft. of countertop area.
- Do not connect lights to the small appliance receptacle circuits.
- Install at least one wall or ceiling-mounted light fixture.

Bathrooms
- Install a separate 20-amp GFCI-protected circuit only for bathroom receptacles.
- Ground switches in bathrooms.
- Install at least one receptacle not more than 36" from each sink.
- Install at least one ceiling- or wall-mounted light fixture.

Utility/Laundry Rooms
- Install a separate 20-amp circuit for a washing machine.
- Install approved conduit for wiring in unfinished rooms.
- Use GFCI-protected circuits for 120-volt and 240-volt receptacles.

Living, Entertainment, Bedrooms
- Install at least one 15- or 20-amp lighting/receptacle circuit for each 600 sq. ft. of living space.
- Install a dedicated circuit for each permanent appliance, such as an air conditioner or group of electric baseboard heaters.
- Use electrical boxes listed and labeled to support ceiling fans.
- Space receptacles on walls in living and sleeping rooms a maximum of 12 feet apart.
- Check with your local electrical inspector about requirements for installing smoke alarms and carbon monoxide alarms during remodeling.

Outdoors
- Check for underground utilities before digging.
- Use UF cable or other wiring approved for wet locations for outdoor wiring.
- Run cable and wires in approved conduit, as required by local code.
- Install in-use rated weatherproof receptacle covers.
- Bury cables and wires run in conduit at least 18" deep; cable not in conduit must be buried at least 24" deep.
- Use weatherproof electrical boxes with watertight covers.
- Install GFCI-protected circuits for receptacles.
- Support boxes that are not attached to a building and that contain switches or receptacles using at least two pieces of conduit. Secure the conduit not more than 18 feet from the box. Locate the box at least 12" above the ground.

Stairs/Hallways
- Use three-way switches at the top and bottom on stairways with six risers or more.
- Include receptacles in any hallway that is 10 feet long or longer.
- Position stairway lights so each step and landing is illuminated.

Prepare for Inspections

Electrical inspectors who issue the permit for your wiring project will also visit your home to review the work. Make sure to allow time for these inspections as you plan the project. For most projects, inspectors make two visits.

The first inspection, called the rough-in, is done after the cables are run between the boxes but before the insulation, wallboard, switches, and fixtures are installed. The second inspection, called the final, is done after the walls and ceilings are finished and all electrical connections are made.

When preparing for the rough-in inspection, make sure the area is neat. Sweep up sawdust and clean up any pieces of scrap wire or cable insulation. Before inspecting the boxes and cables, inspectors will check to make sure all plumbing and other mechanical work is completed. Some electrical inspectors will ask to see your building and plumbing permits.

At the final inspection, inspectors check random boxes to make sure the wire connections are correct. If they see good workmanship at the selected boxes, the inspection will be over quickly. However, if they spot a problem, inspectors may choose to inspect every connection.

Inspectors have busy schedules, so it is a good idea to arrange for an inspection several days in advance. In addition to basic compliance with code, inspectors expect your work to meet their own standards for quality. When you apply for a permit, make sure you understand what the inspectors will look for during inspections.

You cannot put new circuits into use legally until an inspector approves them at the final inspection. If you have planned carefully and done your work well, electrical inspections are routine visits that give you confidence in your own skills.

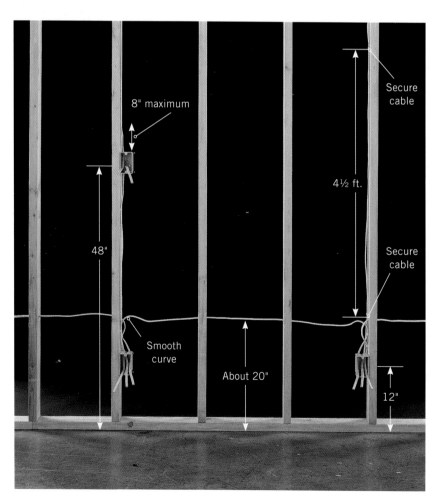

8" maximum

Secure cable

48"

4½ ft.

Secure cable

Smooth curve

About 20"

12"

Inspectors may measure to see that electrical boxes are mounted at consistent heights. Height may not be dictated by code, but consistency is a sign of good workmanship. Measured from the center of the boxes, receptacles in living areas typically are located 12" above the finished floor and switches at 48". For special circumstances, inspectors allow you to alter these measurements. For example, you can install switches at 36" above the floor in a child's bedroom, or set receptacles at 24" to make them more convenient for someone using a wheelchair.

Inspectors will check cables to see that they are secured within 8" of each box and every 4½ ft. thereafter when they run along studs. When bending cables, form the wire in a smooth curve. Do not crimp cables sharply or install them diagonally between framing members. Some inspectors specify that cables running between receptacle boxes should be about 20" above the floor.

What Inspectors Look For

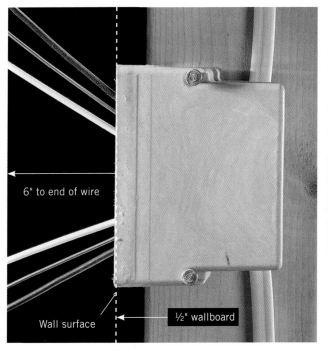

6" to end of wire

Wall surface

½" wallboard

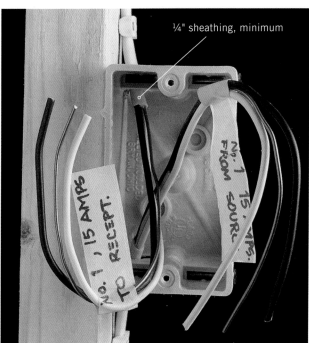

¼" sheathing, minimum

No. 1 , 15 AMPS.
FROM SOURCE

No. 1 , 15 AMPS
TO RECEPT.

Electrical box faces should extend past the front of framing members so the boxes will be flush with finished walls (left). Inspectors will check to see that all boxes are large enough for the wires they contain. Cables should be cut and stripped back so that at least 3" of usable length extends past the front of the box and so that at least ¼" of sheathing reaches into the box (right). Label all cables to show which circuits they serve: inspectors recognize this as a mark of careful work. The labels also simplify the final hookups after the wallboard is installed.

IS YOUR RECEPTACLE SPACING CORRECT?

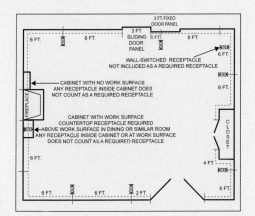

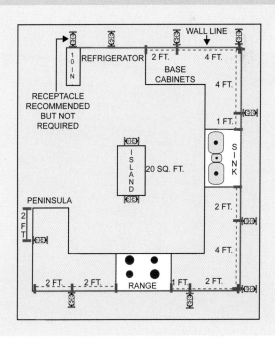

Example of receptacle spacing requirements in a typical room. Measure receptacle spacing distance along the wall line. Install receptacles along partial height walls and along balcony guards in lofts and similar areas.

Example of countertop receptacle spacing in a typical kitchen (right).

Evaluate Electrical Loads

Before drawing a plan and applying for a permit, make sure your home's electrical service provides enough power to handle the added load of the new circuits. In a safe wiring system, the current drawn by fixtures and appliances never exceeds the main service capacity.

To estimate electrical loads, use whatever method is recommended by your electrical inspector. Include the load for all existing wiring as well as that for proposed new wiring when making your estimation.

Most of the light fixtures and plug-in appliances in your home are evaluated as part of general allowances for basic lighting/receptacle circuits and small-appliance circuits. However, appliances that are permanently installed usually require their own dedicated circuits. The electrical loads for these appliances are added in separately when evaluating wiring.

If your evaluation shows that the load exceeds the main service capacity, you must have an electrician upgrade the main service before you can install new wiring. An electrical service upgrade is a worthwhile investment that improves the value of your home and provides plenty of power for present and future wiring projects.

 AMPERAGE

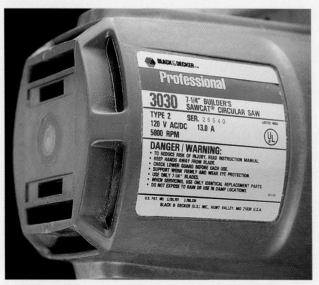

AMPS × VOLTS	TOTAL CAPACITY	SAFE CAPACITY
15 A × 120 V =	1,800 watts	1,440 watts
20 A × 120 V =	2,400 watts	1,920 watts
25 A × 120 V =	3,000 watts	2,400 watts
30 A × 120 V =	3,600 watts	2,880 watts
20 A × 240 V =	4,800 watts	3,840 watts
30 A × 240 V =	7,200 watts	5,760 watts

Amperage rating can be used to find the wattage of an appliance. Multiply the amperage by the voltage of the circuit. For example, a 13-amp, 120-volt circular saw is rated for 1,560 watts.

Calculating Loads

Add 1,500 watts for each small appliance circuit required by the local electrical code. In most communities, three such circuits are required—two in the kitchen and one for the laundry—for a total of 4,500 watts. No further calculations are needed for appliances that plug into small-appliance or basic lighting/receptacle circuits.

Nameplate

If the nameplate gives the rating in kilowatts, find the watts by multiplying kilowatts times 1,000. If an appliance lists only amps, find watts by multiplying the amps times the voltage— either 120 or 240 volts.

FIXED DEVICES

Do not connect one or more fixed devices that in total exceed 50 percent of a multiple outlet branch circuit's amperage rating. Fixed devices do not include light fixtures. This means that that all fixed devices (such as a permanently wired disposal or hot water circulating pump) on a multiple outlet branch circuit may not exceed 7.5 amps (about 900 watts) on a 15-amp multiple outlet branch circuit and may not exceed 10 amps (about 1,200 watts) on a 20-amp multiple outlet branch circuit.

Air-conditioning and heating appliances are not used at the same time, so figure in only the larger of these two numbers when evaluating your home's electrical load.

Locating Wattage

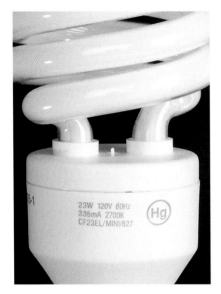

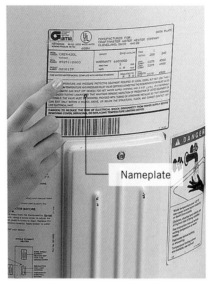

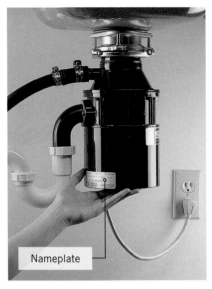

Light bulb wattage ratings are printed on the top of the bulb. If a light fixture has more than one bulb, remember to add the wattages of all the bulbs to find the total wattage of the fixture.

Electric water heaters are permanent appliances that require their own dedicated 30-amp, 240-volt circuits. Most water heaters are rated between 3,500 and 4,500 watts. If the nameplate lists several wattage ratings, use the one labeled "Total Connected Wattage" when figuring electrical loads.

Food disposers are considered permanent appliances and may require their own dedicated 15-amp, 120-volt circuits. Most disposers are rated between 500 and 900 watts.

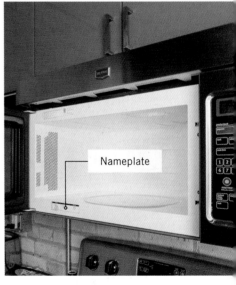

Dishwashers installed permanently under a countertop may need a dedicated 15-amp, 120-volt circuit. Dishwasher ratings are usually between 1,000 and 1,500 watts. Portable dishwashers are regarded as part of small appliance circuits and are not added in when figuring loads.

Electric ranges can be rated for as little as 3,000 watts or as much as 12,000 watts. They require dedicated 120/240-volt circuits. Find the exact wattage rating by reading the nameplate found inside the oven door or on the back of the unit.

Installed microwave ovens are regarded as permanent appliances. Add in its wattage rating when calculating loads. The nameplate is found on the back of the cabinet or inside the front door. Most microwave ovens are rated between 500 and 1,200 watts. A permanently installed microwave should be on a dedicated 20-amp, 120-volt circuit.

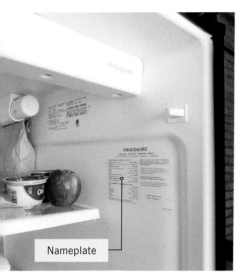

Freezers are appliances that may need a dedicated 15- or 20-amp, 120-volt circuits. Freezer ratings are usually between 240 and 480 watts. But combination refrigerator-freezers are plugged into small appliance circuits and do not need their own dedicated circuits. The nameplate for a freezer is found inside the door or on the back of the unit, just below the door seal.

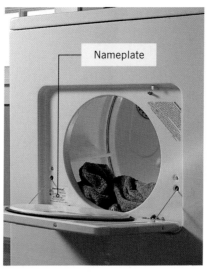

Electric clothes dryers are permanent appliances that need dedicated 30-amp, 120/240-volt circuits. The wattage rating is printed on the nameplate inside the dryer door. Use 5,000 watts as a minimum, regardless of the printed rating. Washing machines and gas-heat clothes dryers with electric tumbler motors do not need dedicated circuits. They plug into the 20-amp small-appliance circuit in the laundry room.

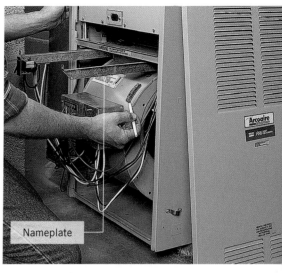

Forced-air furnaces and heat pump air handlers have electric fans and are considered permanent appliances. They require dedicated 15-amp, 120-volt circuits. Include the fan wattage rating, printed on a nameplate inside the control panel, when figuring wattage loads for heating. You should also include the wattage rating for heat pump backup heating coils.

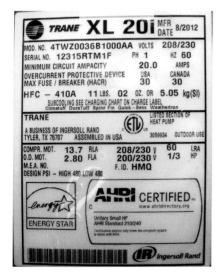

A central air conditioner requires a dedicated 240-volt circuit. Estimate normal wattage use by multiplying the RLA (rated load amps) by 240 (volts).

Window air conditioners may be considered permanent appliances if they are connected to a dedicated circuit. Through-wall air conditioners are considered permanent appliances. The wattage rating, which can range from 500 to 2,000 watts, is found on the nameplate located inside the front grill. Include permanently installed through-wall air conditioners and window air conditioners that are connected to a dedicated circuit in your evaluation.

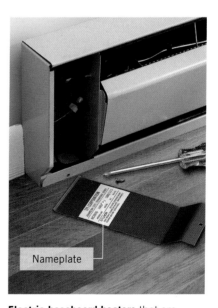

Electric baseboard heaters that are permanently installed require a dedicated circuit and must be figured into the load calculations. Use the maximum wattage rating printed inside the cover. In general, 240-volt baseboard-type heaters are rated for 180 to 250 watts for each linear foot.

Draw a Diagram & Obtain a Permit

Drawing a wiring diagram is the last step in planning a circuit installation. A detailed wiring diagram helps you get a permit, makes it easy to create a list of materials, and serves as a guide for laying out circuits and installing cables and fixtures. Use the circuit maps on pages 32–47 as a guide for planning wiring configurations and cable runs. Bring the diagram and materials list when you visit electrical inspectors to apply for a permit.

Never install new wiring without following your community's permit and inspection procedure. A permit is not expensive, and it ensures that your work will be reviewed by a qualified inspector. If you install new wiring without the proper permit, an accident or fire traced to faulty wiring could cause your insurance company to discontinue your policy and can hurt the resale value of your home.

When electrical inspectors look over your wiring diagram, they will ask questions to see if you have a basic understanding of the electrical code and fundamental wiring skills. Some inspectors ask these questions informally, while others give a short written test. Inspectors may allow you to do some, but not all, of the work. For example, they may ask that all final circuit connections at the circuit breaker panel be made by a licensed electrician, while allowing you to do all other work.

A few communities allow you to install wiring only when supervised by an electrician. This means you can still install your own wiring but must hire an electrician to apply for the work permit and to check your work before inspectors review it. The electrician is held responsible for the quality of the job.

Remember that it is the inspectors' responsibility to help you do a safe and professional job. Feel free to call them with questions about wiring techniques or materials.

A detailed wiring diagram and a list of materials is required before electrical inspectors will issue a work permit. If blueprints exist for the space you are remodeling, start your electrical diagram by tracing the wall outlines from the blueprint. Use standard electrical symbols (next page) to clearly show all the receptacles, switches, light fixtures, and permanent appliances. Make a copy of the symbol key and attach it to the wiring diagram for the inspectors' convenience. Show each cable run, and label its wire size and circuit amperage.

How to Draw a Wiring Plan

Draw a scaled diagram of the space you will be wiring, showing walls, doors, windows, plumbing pipes and fixtures, and heating and cooling ducts. Find the floor space by multiplying room length by width, and indicate this on the diagram.

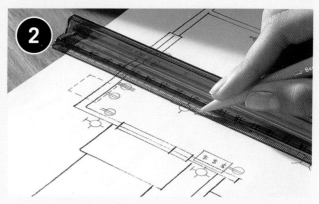

Mark the location of all switches, receptacles, light fixtures, and permanent appliances, using the electrical symbols shown below. Where you locate these devices along the cable run determines how they are wired. Use the circuit maps on pages 32–47 as a guide for drawing wiring diagrams.

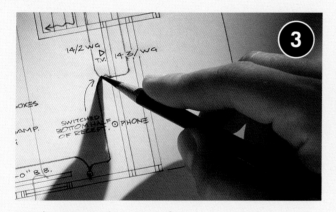

Draw in cable runs between devices. Indicate cable size and type and the amperage of the circuits. Use a different-colored pencil for each circuit.

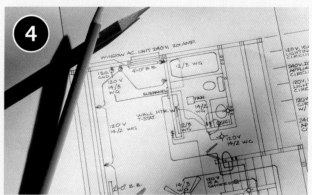

Identify the wattages for light fixtures and permanent appliances and the type and size of each electrical box. On another sheet of paper, make a detailed list of all materials you will use.

ELECTRICAL SYMBOL KEY (COPY THIS KEY AND ATTACH IT TO YOUR WIRING PLAN)

240-volt receptacle	Switched receptacle	J — Junction box	CF — Ceiling fan
Isolated ground receptacle *	Weatherproof receptacle WP	S — Ceiling pull switch	D — Electric door opener
Duplex receptacle	S_TH — Thermostat	Surface-mounted light fixture	BT — Low-voltage transformer
240-volt dryer receptacle D	S_P — Pilot-light switch	R — Recessed light fixture	TV — Television jack
Singleplex receptacle	S — Single-pole switch	Fluorescent light fixture	Data Port
Fourplex receptacle	S_T — Timer switch	Wall-mounted light fixture	D — Smoke alarm
GFCI duplex receptacle GFCI	S_3 — Three-way switch	WP — Weatherproof light fixture	VF — Exhaust fan

Wiring a Room Addition

The photo below shows the circuits you would likely want to install in a large room addition. This example shows the framing and wiring of an unfinished attic converted to an office or entertainment room with a bathroom. This room includes a subpanel and five new circuits plus telephone and cable-TV lines.

A wiring project of this sort is a potentially complicated undertaking that can be made simpler by breaking the project into convenient steps, and finishing one step before moving on to the next. Turn to pages 24–25 to see this project represented as a wiring diagram.

14/2 cable

Exhaust fan

Circuit breaker subpanel

Vanity light fixture

GFCI receptacle

12/2 cable

12/2 cable

Time & light fixture switch

14/3 cable

Blower heater

10/3 cable

INDIVIDUAL CIRCUITS

#1: Bathroom circuit. This 20-amp dedicated circuit supplies power to bathroom lights and fans, as well as receptacles that must be GFCI-protected at the box or at the receptacle. As with small appliance circuits in the kitchen, you may not tap into this circuit to feed any additional loads.

#2: Computer circuit. A 15-amp dedicated circuit with isolated ground is recommended, but an individual branch circuit is all that is required by most codes.

Circuit breaker subpanel receives power through a 10-gauge, three-wire (plus ground) feeder cable connected to a 30-amp, 240-volt circuit breaker at the main circuit breaker panel. Larger room additions may require a 60-or 100-amp feeder circuit breaker.

#3: Air-conditioner circuit. This is a 20-amp, 240-volt dedicated circuit. In cooler climates, or in a smaller room, you may need an air conditioner and circuit rated for only 120 volts.

#4: Basic lighting/receptacle circuit. This 15-amp, 120-volt circuit supplies power to most of the fixtures in the bedroom and study areas.

#5: Heater circuit. This 20-amp, 240-volt circuit supplies power to the bathroom blower-heater and to the baseboard heaters. Depending on the size of your room and the wattage rating of the baseboard heaters, you may need a 30-amp, 240-volt heating circuit.

Computer data outlet is usually wired with Cat 6 (or better) cable that runs to the internet provider's exterior interface box. Each outlet should have a dedicated (home run) cable to the interface box, unless you connect devices (like a printer) in other rooms using cable instead of Wi-Fi.

Cable television outlet is wired with RG59 (or better) 75 ohm coaxial cable that runs to the cable provider's exterior interface box. Each outlet should have a dedicated (home run) cable to the interface box. Note that computer and cable outlets are often combined into one multipurpose outlet.

14/3 cable

14/3 cable

Computer data and cable television outlet

Coaxial cable

12/2 cable

These cables continue through the foreground wall to complete the circuits. This wall has been removed for clarity.

14/2 cable

Diagram View

The diagram below shows the layout of the five circuits and the locations of their receptacles, switches, fixtures, and devices as shown in the photo on the previous pages. The circuits and receptacles are based on the needs of a 400-square-foot space. An inspector will want to see a diagram like this one before issuing a permit. After you've received approval for your addition, the wiring diagram will serve as your guide as you complete your project.

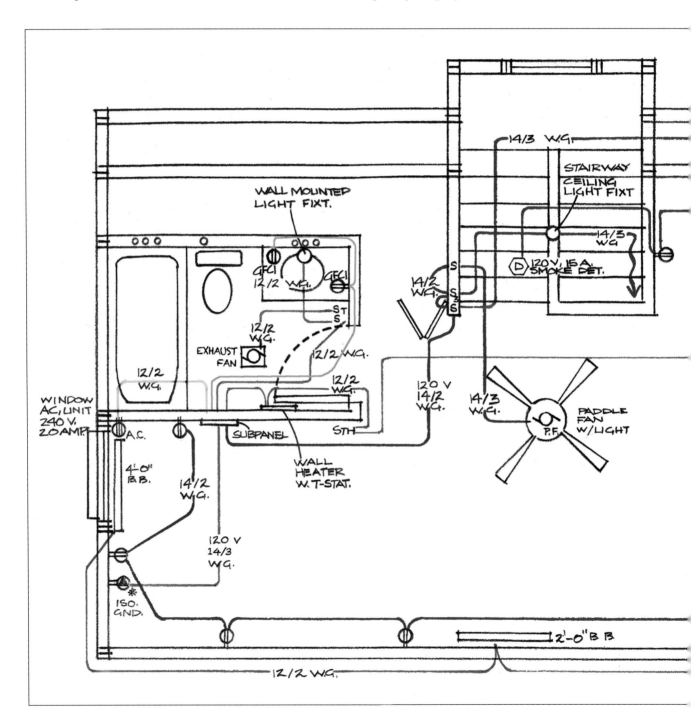

Circuit #1: A 20-amp, 120-volt circuit serving the bathroom. Includes: 12/2 NM cable, double-gang box, timer switch, single-pole switch, 4" × 4" box with single-gang adapter plate, plastic light fixture boxes, vanity light fixture, 20-amp single-pole circuit breaker.

Circuit #2: A 15-amp, 120-volt computer circuit. Includes: 14/3 NM cable, single-gang box, 15-amp receptacle, 15-amp single-pole circuit breaker.

Circuit #3: A 20-amp, 240-volt air-conditioner circuit. Includes: 12/2 NM cable; single-gang box; 20-amp, 240-volt receptacle (singleplex style); 20-amp double-pole circuit.

Circuit #4: A 15-amp, 120-volt basic lighting/receptacle circuit serving most of the fixtures in the bedroom and study areas. Includes: 14/2 and 14/3 NM cable, two double-gang boxes, fan speed-control switch, dimmer switch, single-pole switch, two three-way switches, two plastic light fixture boxes, light fixture for stairway, smoke detector, metal light fixture box with brace bar, ceiling fan with light fixture, 10 single-gang boxes, 4" × 4" box with single-gang adapter plate, 10 duplex receptacles (15-amp), 15-amp single-pole circuit breaker.

Circuit #5: A 20-amp, 240-volt circuit that supplies power to three baseboard heaters controlled by a wall thermostat and to a bathroom blower-heater controlled by a built-in thermostat. Includes: 12/2 NM cable, 750-watt blower heater, single-gang box, line-voltage thermostat, three baseboard heaters, 20-amp double-pole circuit breaker.

TV **Cable television outlet** is wired with RG59 (or better) 75 ohm coaxial cable that runs to the cable provider's exterior interface box. Each outlet should have a dedicated (home run) cable to the interface box. Note that computer and cable outlets are often combined into one multipurpose outlet.

▽ **Computer data outlet** is usually wired with Cat 6 (or better) cable that runs to the internet provider's exterior interface box. Each outlet should have a dedicated (home run) cable to the interface box, unless you connect devices (like a printer) in other rooms using cable instead of Wi-Fi.

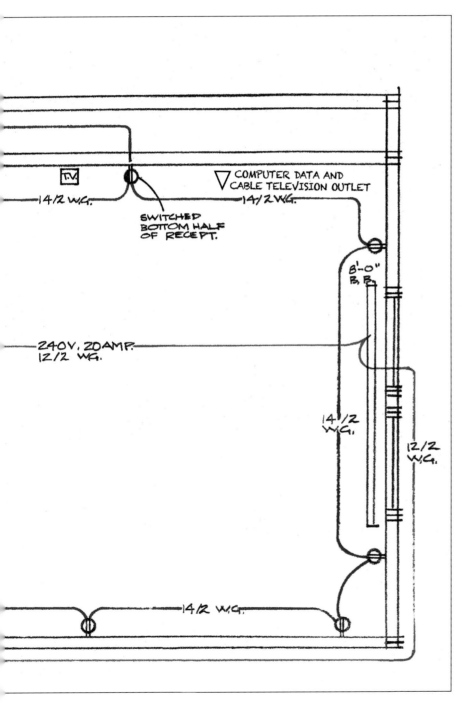

Wiring a Kitchen

14/3 cable

12/3 cable

12/2 cable

12/2 cable

6/3 cable

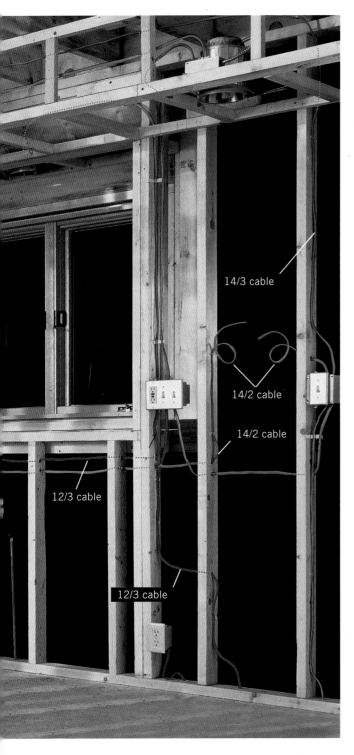

14/3 cable

14/2 cable

14/2 cable

12/3 cable

12/3 cable

The photo at left shows the circuits you would probably want to install in a total kitchen remodel. Kitchens require a wide range of electrical services, from simple 15-amp lighting circuits to 120/240, 50-amp appliance circuits. This kitchen example has six circuits, including a dedicated circuit for a dishwasher and food disposer. Some codes allow the disposer and dishwasher to share a single circuit.

All rough carpentry and plumbing should be in place before beginning any electrical work. As always, divide a project of this scale into manageable steps, and finish one step before moving on. Turn to pages 28–29 to see this project represented as a wiring diagram.

INDIVIDUAL CIRCUITS

■ **#1 & #2: Small-appliance circuits.** Two 20-amp, 120-volt circuits supply power to countertop dining area receptacles. All general-use receptacles must be on these circuits. One 12/3 cable fed by a 20-amp double-pole breaker wires both circuits. Other circuits may service the area, such as a dedicated refrigerator circuit.

■ **#3: Range circuit.** A 50-amp, 120/240-volt dedicated circuit supplies power to the range. It is wired with 6/3 copper cable.

■ **#4: Microwave circuit.** It is wired with 12/2 cable.

▨ **#5: Food disposer/dishwasher circuit.** A dedicated 20-amp circuit supplies power for the dishwasher and the food disposer. It is wired with 12/2 cable. This circuit often terminates in one box with one half of the receptacle serving the dishwasher and the other half serving the food disposer. This is a GFCI-protected circuit.

■ **#6: Basic lighting circuit.** A 15-amp, 120-volt circuit powers the ceiling fixture, dining area fixture, recessed fixtures, and undercabinet task lights. 14/2 and 14/3 cables connect the fixtures and switches in the circuit. Each task light has a self-contained switch.

This visual is an example of how kitchen wiring may look. If you are wiring a kitchen, please use the diagram on page 29 for best results.

Diagram View

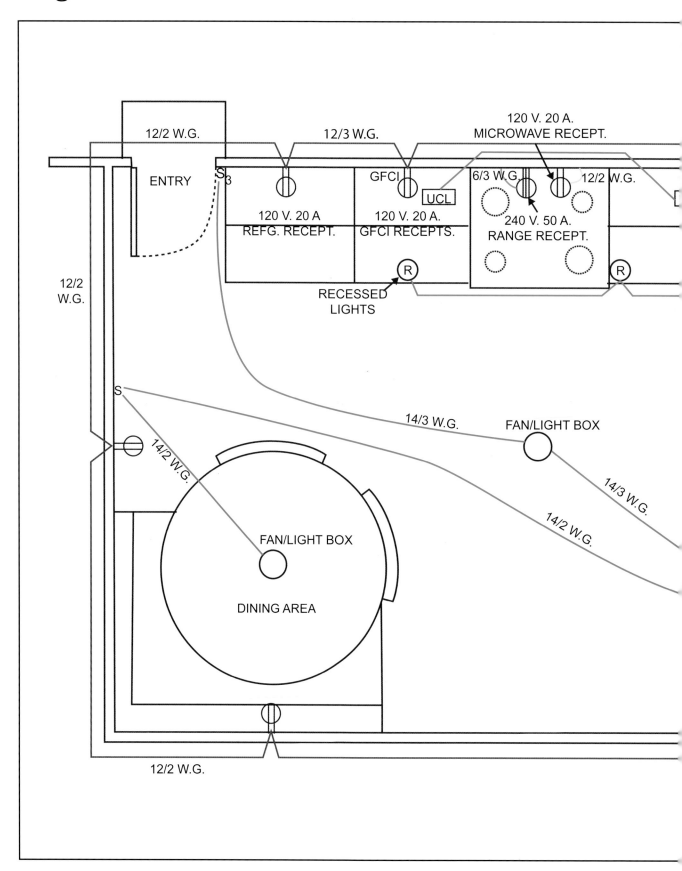

120 V. 20 A.
MICROWAVE RECEPT.

12/2 W.G.

12/3 W.G.

ENTRY

S₃

GFCI

UCL

6/3 W.G.

12/2 W.G.

120 V. 20 A
REFG. RECEPT.

120 V. 20 A.
GFCI RECEPTS.

240 V. 50 A.
RANGE RECEPT.

R

R

12/2
W.G.

RECESSED
LIGHTS

S

14/3 W.G.

FAN/LIGHT BOX

14/3 W.G.

14/2 W.G.

14/2 W.G.

FAN/LIGHT BOX

DINING AREA

12/2 W.G.

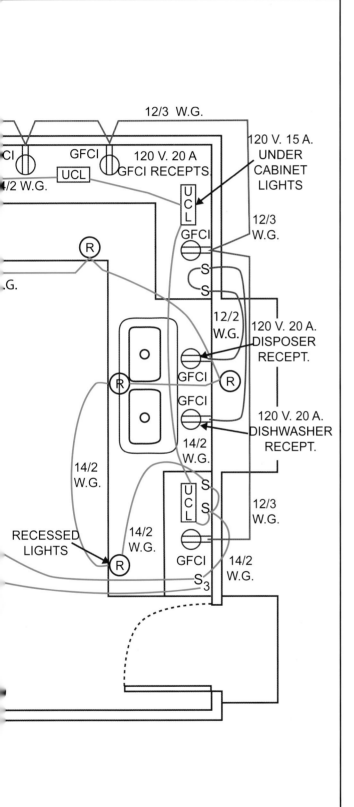

The diagram at left shows the layout of the seven circuits and the locations of their receptacles, switches, fixtures, and devices as shown in the photo on the previous pages. The circuits and receptacles are based on the needs of a 175-sq.-ft. space kitchen. An inspector will want to see a diagram like this one before issuing a permit. After you've received approval for your addition, the wiring diagram will serve as your guide as you complete your project.

Circuits #1 & #2: Two 20-amp, 120-volt circuits supply power to countertop dining area receptacles. All general-use receptacles must be on these circuits. One 12/3 cable fed by a 20-amp double-pole breaker wires both circuits. Other circuits may service the area, such as a dedicated refrigerator circuit.

Circuit #3: A 50-amp, 120/240-volt dedicated circuit supplies power to the range. It is wired with 6/3 copper cable.

Circuit #4: It is wired with 12/2 cable.

Circuit #5: A dedicated 20-amp circuit supplies power for the dishwasher and the food disposer. It is wired with 12/2 cable. This circuit often terminates in one box with one half of the receptacle serving the dishwasher and the other half serving the food disposer. This is a GFCI-protected circuit.

Circuit #6: A 15-amp, 120-volt circuit powers the ceiling fixture, dining area fixture, recessed fixtures, and undercabinet task lights. 14/2 and 14/3 cables connect the fixtures and switches in the circuit. Each task light has a self-contained switch.

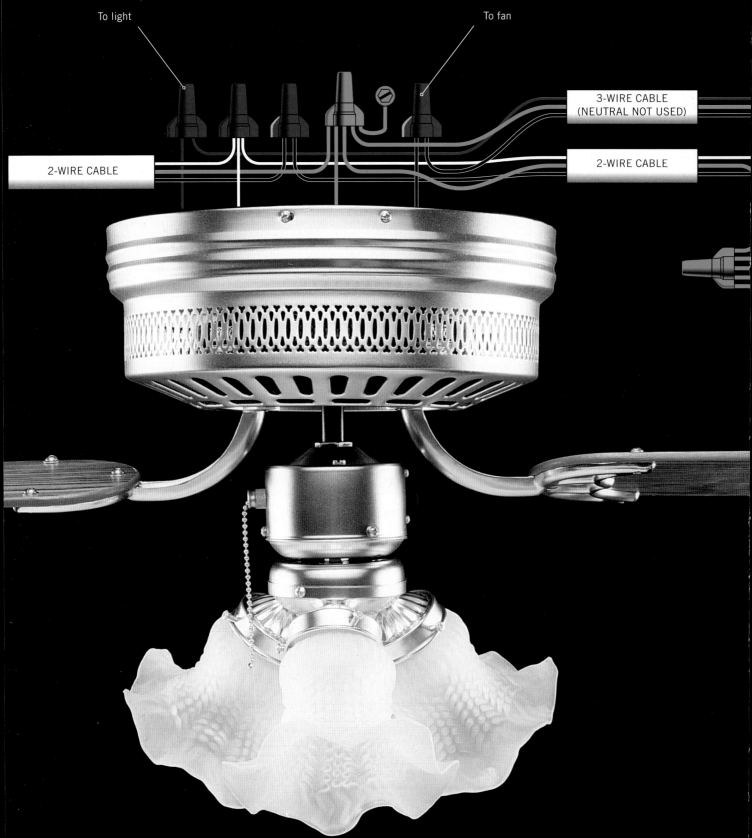

To light

To fan

3-WIRE CABLE
(NEUTRAL NOT USED)

2-WIRE CABLE

2-WIRE CABLE

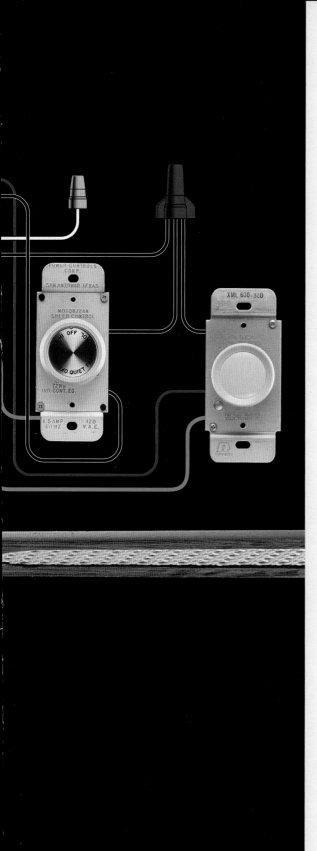

Special Section: Circuit Maps

The circuit maps on the following pages show the most common wiring variations for typical electrical devices. Most new wiring you install will match one or more of the maps shown. Find the maps that match your situation and use them to plan your circuit layouts.

The 120-volt circuits shown on the following pages are wired for 15 amps using 14-gauge wire and receptacles rated at 15 amps. If you are installing a 20-amp circuit, substitute 12-gauge cables and use receptacles rated for 15 or 20 amps.

In configurations where a white wire serves as a hot wire instead of a neutral, both ends of the wire are coded with black tape to identify it as hot. In addition, each of the circuit maps shows a box grounding screw. This grounding screw is required in all metal boxes, but plastic electrical boxes do not need to be grounded.

You should remember two code requirements when wiring switches. (1) Provide a neutral wire at every switch box. This may require using 3-wire cable or two 2-wire cables where you may have used one 2-wire cable in the past. (2) Use a black or red wire to supply power from a 3-way or a 4-way switch to a light or switched receptacle.

NOTE: For clarity, all grounding conductors in the circuit maps are colored green. In practice, the grounding wires inside sheathed cables usually are bare copper.

Common Household Circuits

1. 120-Volt Duplex Receptacles Wired in Sequence

Use this layout to link any number of duplex receptacles in a basic lighting/receptacle circuit. The last receptacle in the cable run is connected like the receptacle shown at the right side of the circuit map below. All other receptacles are wired like the receptacle shown on the left side. This configuration or layout requires two-wire cables.

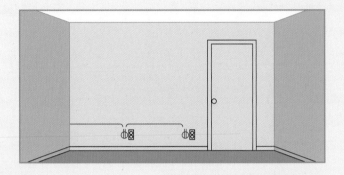

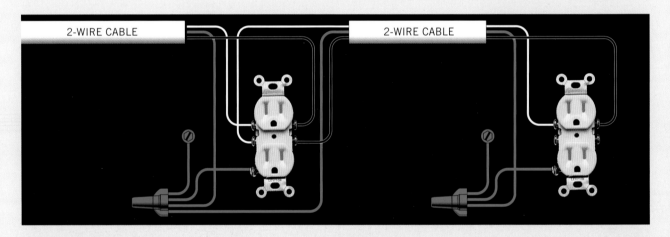

2-WIRE CABLE · 2-WIRE CABLE

2. GFCI Receptacles (Single-Location Protection)

Use this layout when receptacles are within 6 ft. of a water source, such as those in kitchens and bathrooms. To prevent nuisance tripping caused by normal power surges, GFCIs should be connected only at the line screw terminal so they protect a single location, not the fixtures on the load side of the circuit. Requires two-wire cables. Where a GFCI must protect other fixtures, use circuit map 3. Remember that bathroom receptacles should be on a dedicated 20-amp circuit and that all bathroom receptacles must be GFCI-protected.

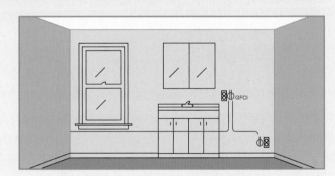

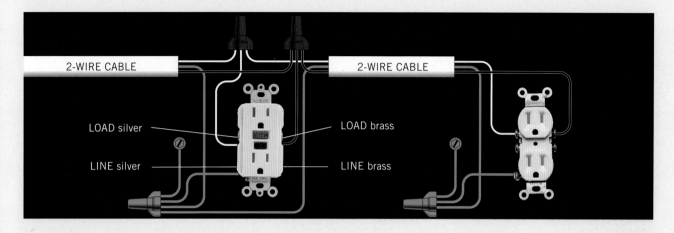

2-WIRE CABLE   2-WIRE CABLE

LOAD silver   LOAD brass

LINE silver   LINE brass

3. GFCI Receptacle, Switch & Light Fixture (Wired for Multiple-Location Protection)

In some locations, such as an outdoor circuit, it is a good idea to connect a GFCI receptacle so it also provides shock protection to the wires and fixtures that continue to the end of the circuit. Wires from the power source are connected to the line screw terminals; outgoing wires are connected to load screws. Requires two-wire cables.

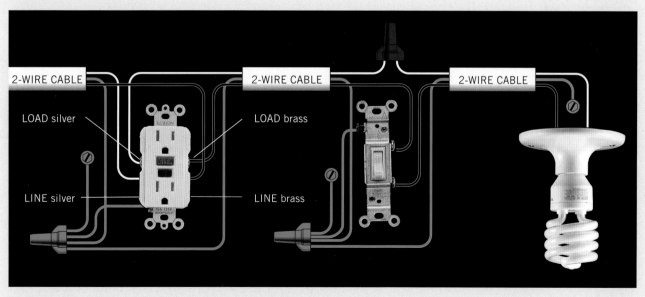

2-WIRE CABLE   2-WIRE CABLE   2-WIRE CABLE

LOAD silver   LOAD brass

LINE silver   LINE brass

4. Single-Pole Switch & Light Fixture
(Light Fixture at End of Cable Run)

Use this layout for light fixtures in basic lighting/ receptacle circuits throughout the home. It is often used as an extension to a series of receptacles (circuit map 1). Requires two-wire cables.

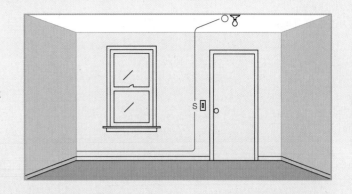

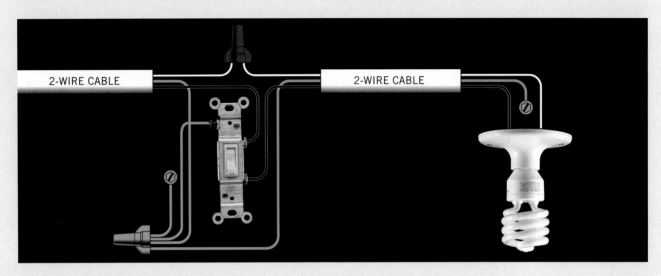

5. Single-Pole Switch & Light Fixture
(Switch at End of Cable Run)

Use this layout, sometimes called a switch loop, where it is more practical to locate a switch at the end of the cable run. In the last length 3-wire cable is used to make a hot conductor available in each direction. Requires two-wire and three-wire cables.

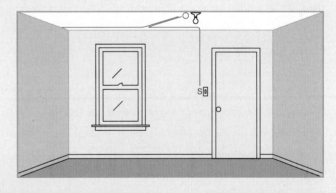

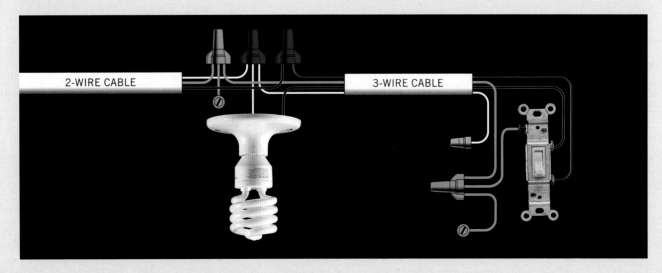

6. Single-Pole Switch & Two Light Fixtures
(Switch Between Light Fixtures, Light at Start of Cable Run)

Use this layout when you need to control two fixtures from one single-pole switch and the switch is between the two lights in the cable run. Power feeds to one of the lights. Requires two-wire and three-wire cables.

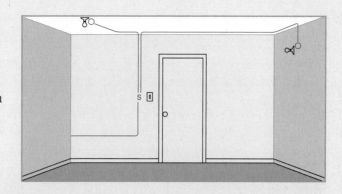

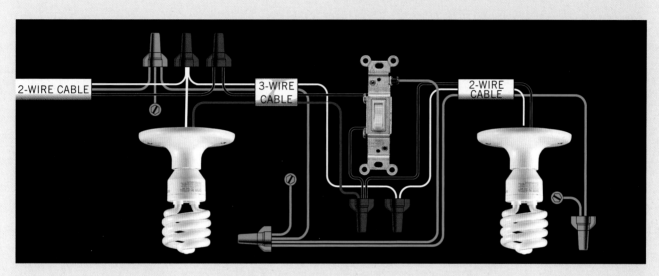

7. Single-Pole Switch & Light Fixture, Duplex Receptacle
(Switch at Start of Cable Run)

Use this layout to continue a circuit past a switched light fixture to one or more duplex receptacles. To add multiple receptacles to the circuit, see circuit map 1. Requires two-wire and three-wire cables.

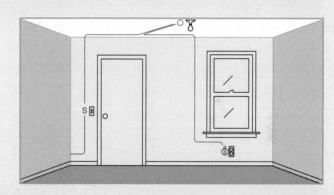

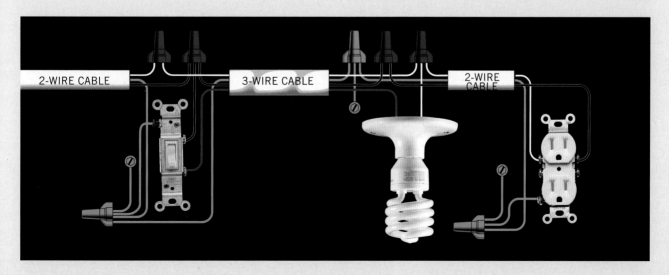

8. Switch-Controlled Split Receptacle, Duplex Receptacle (Switch at Start of Cable Run)

This layout lets you use a wall switch to control a lamp plugged into a wall receptacle. This configuration is required by code for any room that does not have a switch-controlled wall or ceiling fixture. Only the bottom half of the first receptacle is controlled by the wall switch; the top half of the receptacle and all additional receptacles on the circuit are always hot. Requires two-wire and three-wire cables. Some electricians help people identify switched receptacles by installing them upside down.

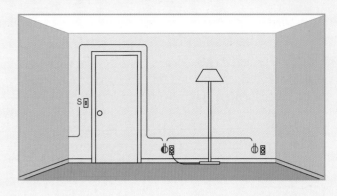

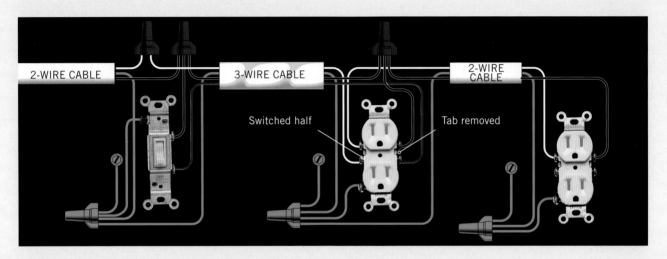

2-WIRE CABLE

3-WIRE CABLE

2-WIRE CABLE

Switched half

Tab removed

9. Switch-Controlled Split Receptacle (Switch at End of Cable Run)

Use this switch loop layout to control a split receptacle (see circuit map 7) from an end-of-run circuit location. The bottom half of the receptacle is controlled by the wall switch, while the top half is always hot. Requires two-wire and three-wire cable. Some electricians help people identify switched receptacles by installing them upside down.

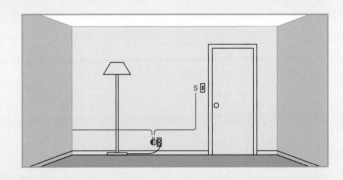

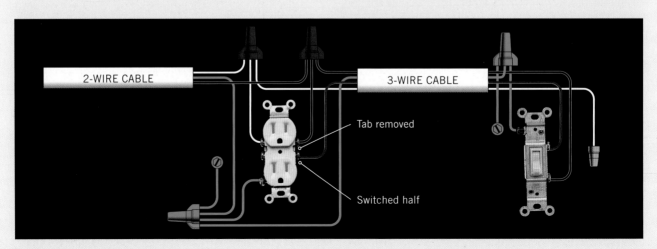

2-WIRE CABLE

3-WIRE CABLE

Tab removed

Switched half

10. Switch-Controlled Split Receptacle, Duplex Receptacle
(Split Receptacle at Start of Run)

Use this variation of circuit map 7 where it is more practical to locate a switch-controlled receptacle at the start of a cable run. Only the bottom half of the first receptacle is controlled by the wall switch; the top half of the receptacle, and all other receptacles on the circuit, are always hot. Requires two-wire and three-wire cables. Some electricians help people identify switched receptacles by installing them upside down.

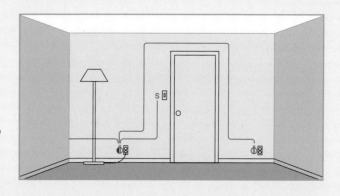

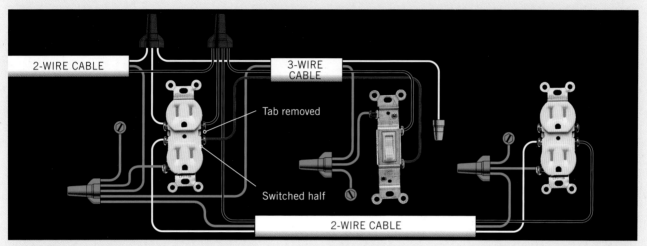

2-WIRE CABLE

3-WIRE CABLE

Tab removed

Switched half

2-WIRE CABLE

11. Double Receptacle Circuit with Shared Neutral Wire
(Receptacles Alternate Circuits)

This layout features two 120-volt circuits wired with one three-wire cable connected to a double-pole circuit breaker. The black hot wire powers one circuit; the red wire powers the other. The white wire is a shared neutral that serves both circuits. When wired with 12/2 and 12/3 cable and receptacles rated for 20 amps, this layout can be used for the two small-appliance circuits required in a kitchen. Remember to use a GFCI circuit breaker if you use this circuit for kitchen counter top receptacles.

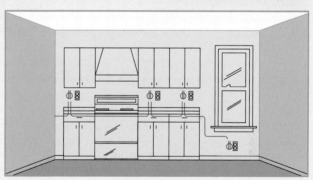

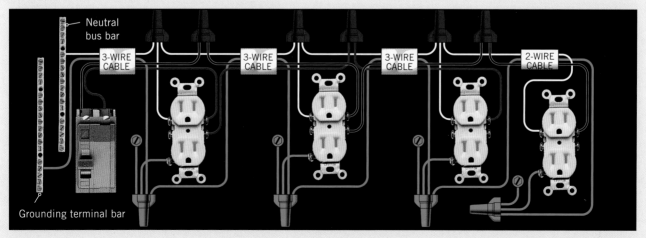

Neutral bus bar

3-WIRE CABLE

3-WIRE CABLE

3-WIRE CABLE

2-WIRE CABLE

Grounding terminal bar

12. Double Receptacle Small-Appliance Circuit with GFCIs & Shared Neutral Wire

Use this layout variation of circuit map 10 to wire a double receptacle circuit when code requires that some of the receptacles be GFCIs. The GFCIs should be wired for single-location protection (see circuit map 2). Requires three-wire and two-wire cables. A GFCI breaker would be another option.

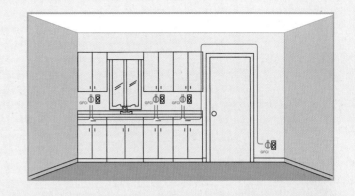

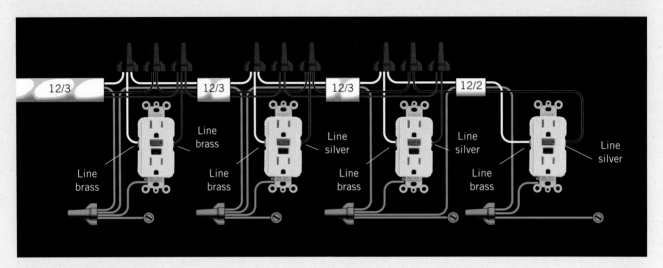

13. Double Receptacle Small-Appliance Circuit with GFCIs & Separate Neutral Wires

If the room layout or local codes do not allow for a shared neutral wire, use this layout instead. The GFCIs should be wired for single-location protection (see circuit map 2). Requires two-wire cable. A GFCI breaker would be another option.

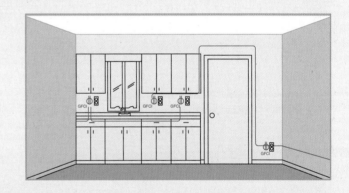

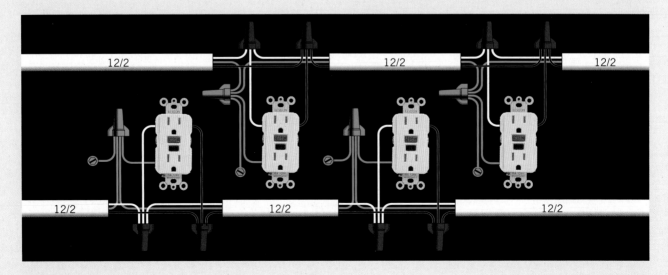

14. 120/240-Volt Range Receptacle

This layout is for a 40- or 50-amp, 120/240-volt dedicated appliance circuit wired with 8/3 or 6/3 cable, as required by code for a large kitchen range. The black and red circuit wires, connected to a double-pole circuit breaker in the circuit breaker panel, each bring 120 volts of power to the setscrew terminals on the receptacle. The white circuit wire attached to the neutral terminal bar in the circuit breaker panel is connected to the neutral setscrew terminal on the receptacle.

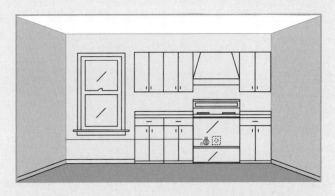

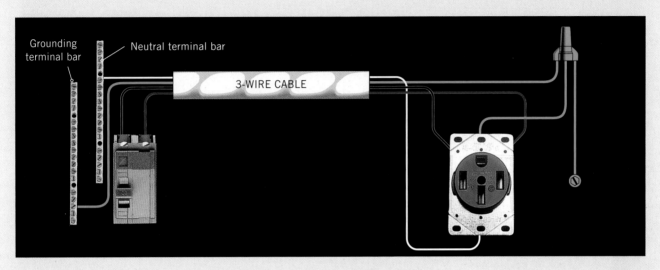

15. 240-Volt Baseboard Heaters, Thermostat

This layout is typical for a series of 240-volt baseboard heaters controlled by a wall thermostat. Except for the last heater in the circuit, all heaters are wired as shown below. The last heater is connected to only one cable. The sizes of the circuit and cables are determined by finding the total wattage of all heaters. Requires two-wire cable.

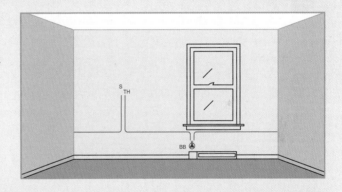

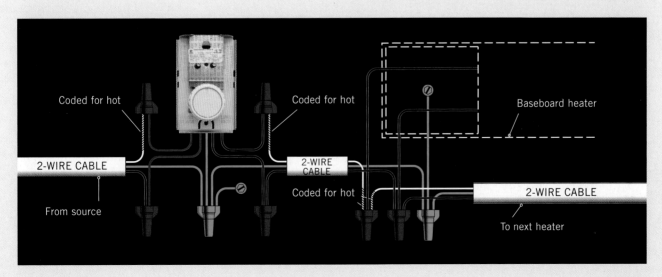

16. Dedicated 120-Volt Computer Circuit, Isolated-Ground Receptacle

This 15-amp isolated-ground circuit provides extra protection against surges and interference that can harm electronics. It uses 14/3 cable with the red wire serving as an extra grounding conductor. The red wire is tagged with green tape for identification. It is connected to the grounding screw on an isolated-ground receptacle and runs back to the grounding terminal bar in the circuit breaker panel without touching any other house wiring.

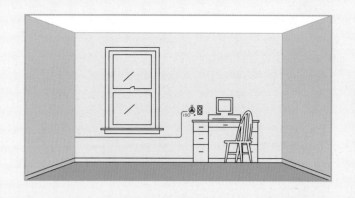

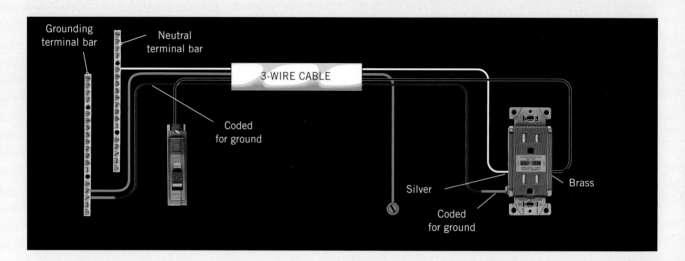

17. 240-Volt Appliance Receptacle

This layout represents a 20-amp, 240-volt dedicated appliance circuit wired with 12/2 cable, as required by code for a large window air conditioner. Receptacles are available in both singleplex (shown) and duplex styles. The black and the white circuit wires connected to a double-pole breaker each bring 120 volts of power to the receptacle (combined, they bring 240 volts). The white wire is tagged with black tape to indicate it is hot.

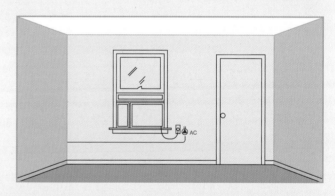

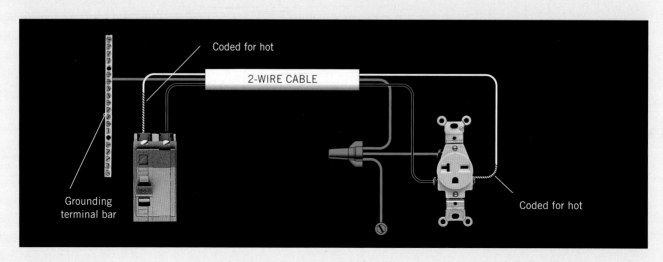

18. Ganged Single-Pole Switches Controlling Separate Light Fixtures

This layout lets you place two switches controlled by the same 120-volt circuit in one double-gang electrical box. A single-feed cable provides power to both switches. A similar layout with two feed cables can be used to place switches from different circuits in the same box. Requires two-wire cable.

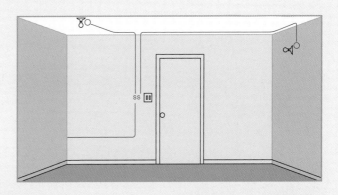

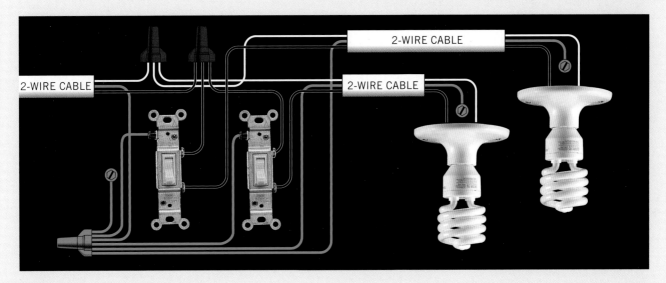

19. Ganged Switches Controlling a Light Fixture and an Exhaust Fan

This layout lets you place two switches controlled by the same 120-volt circuit in one double-gang electrical box. A single-feed cable provides power to both switches. A standard switch controls the light fixture, and a time-delay switch controls the exhaust fan.

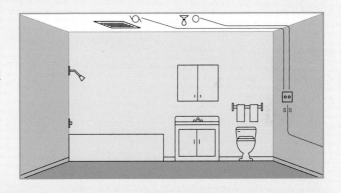

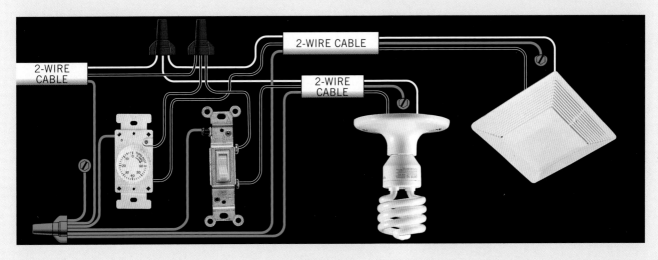

20. Three-Way Switches & Light Fixture
(Fixture Between Switches)

This layout for three-way switches lets you control a light fixture from two locations. Each switch has one common screw terminal and two traveler screws. Circuit wires attached to the traveler screws run between the two switches, and hot wires attached to the common screws bring current from the power source and carry it to the light fixture. Requires parallel runs of 2-wire cable.

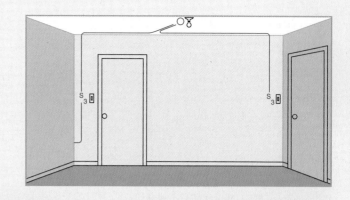

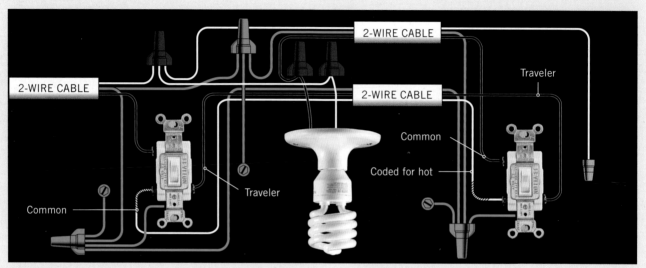

21. Three-Way Switches & Light Fixture
(Fixture at Start of Cable Run)

Use this layout variation of circuit map 19 where it is more convenient to locate the fixture ahead of the three-way switches in the cable run. Requires two-wire and three-wire cables.

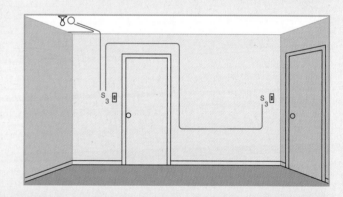

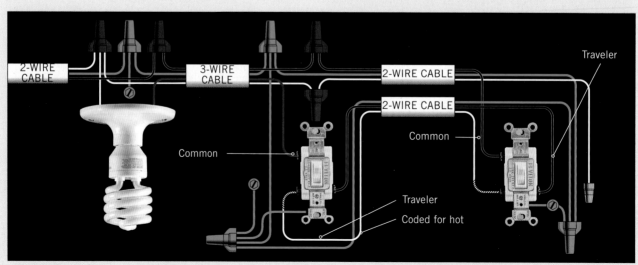

22. Three-Way Switches & Light Fixture (Fixture at End of Cable Run)

This variation of the three-way switch layout (circuit map 20) is used where it is more practical to locate the fixture at the end of the cable run. Requires two-wire and three-wire cables.

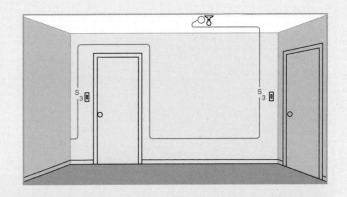

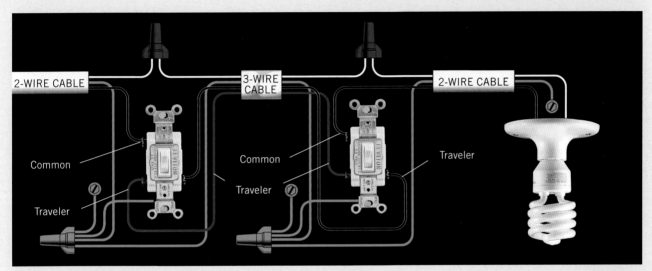

2-WIRE CABLE

3-WIRE CABLE

2-WIRE CABLE

Common

Traveler

Common

Traveler

Traveler

23. Three-Way Switches & Light Fixture with Duplex Receptacle

Use this layout to add a receptacle to a three-way switch configuration (circuit map 21). Requires two-wire and parallel runs of two-wire cables.

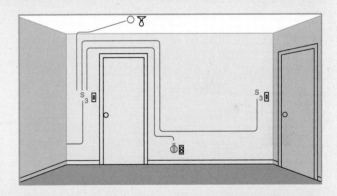

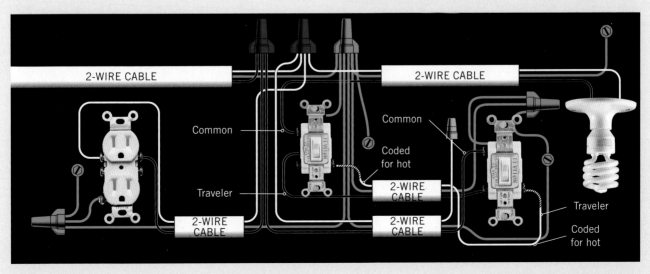

2-WIRE CABLE

2-WIRE CABLE

Common

Common

Coded for hot

2-WIRE CABLE

Traveler

2-WIRE CABLE

2-WIRE CABLE

Traveler

Coded for hot

24. Three-Way Switches & Multiple Light Fixtures (Fixtures Between Switches)

This is a variation of circuit map 20. Use it to place multiple light fixtures between two three-way switches where power comes in at one of the switches. Requires two- and three-wire cable.

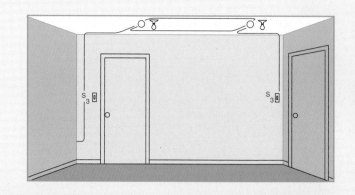

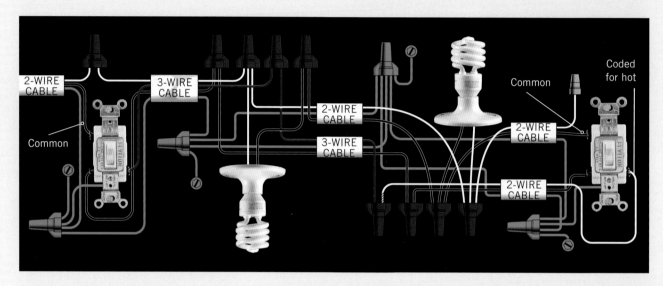

25. Three-Way Switches & Multiple Light Fixtures (Fixtures at Beginning of Run)

This is a variation of circuit map 21. Use it to place multiple light fixtures at the beginning of a run controlled by two three-way switches. Power comes in at the first fixture. Requires two- and three-wire cable.

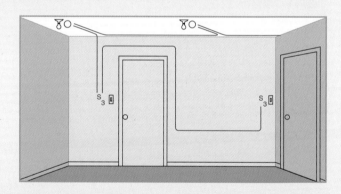

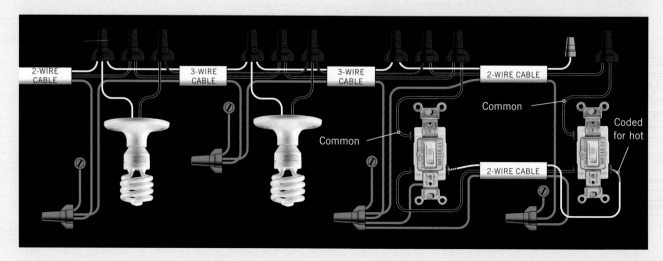

26. Four-Way Switch & Light Fixture (Fixture at Start of Cable Run)

This layout lets you control a light fixture from three locations. The end switches are three-way, and the middle is four-way. A pair of three-wire cables enter the box of the four-way switch. The white and red wires from one cable attach to the top pair of screw terminals (line 1), and the white and red wires from the other cable attach to the bottom screw terminals (line 2). Requires two three-way switches and one four-way switch and two-wire and three-wire cables.

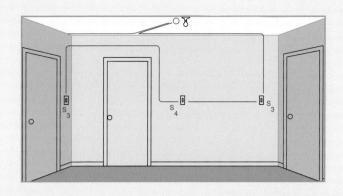

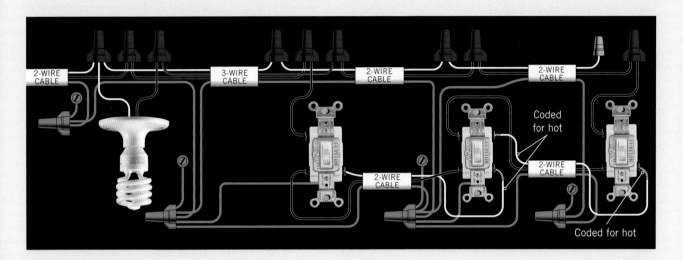

27. Four-Way Switch & Light Fixture (Fixture at End of Cable Run)

Use this layout variation of circuit map 26 where it is more practical to locate the fixture at the end of the cable run. Requires two three-way switches and one four-way switch and two-wire and three-wire cables.

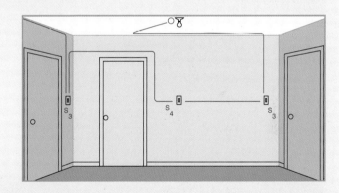

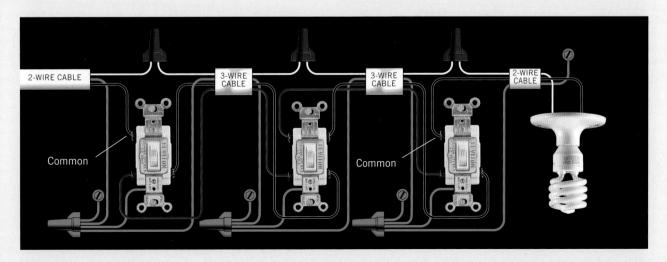

28. Multiple Four-Way Switches
Controlling a Light Fixture

This alternate variation of the four-way switch layout (circuit map 27) is used where three or more switches will control a single fixture. The outer switches are three-way, and the middle are four-way. Requires two three-way switches and two four-way switches and two-wire and three-wire cables.

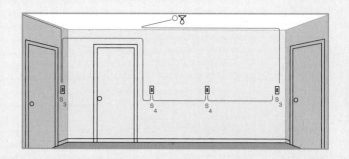

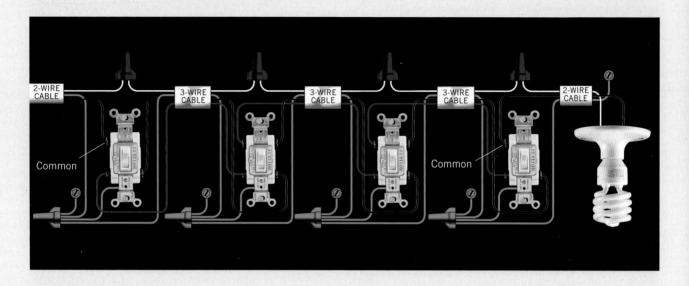

29. Four-Way Switches
& Multiple Light Fixtures

This variation of the four-way switch layout (circuit map 26) is used where two or more fixtures will be controlled from multiple locations in a room. Outer switches are three-way, and the middle switch is a four-way. Requires two three-way switches and one four-way switch and two-wire and three-wire cables.

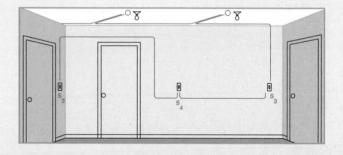

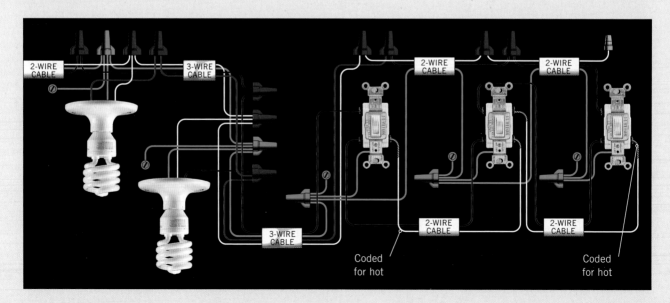

30. Ceiling Fan/Light Fixture Controlled by Ganged Switches
(Fan at End of Cable Run)

This layout is for a combination ceiling fan/light fixture controlled by a speed-control switch and dimmer in a double-gang switch box. Requires two-wire and three-wire cables.

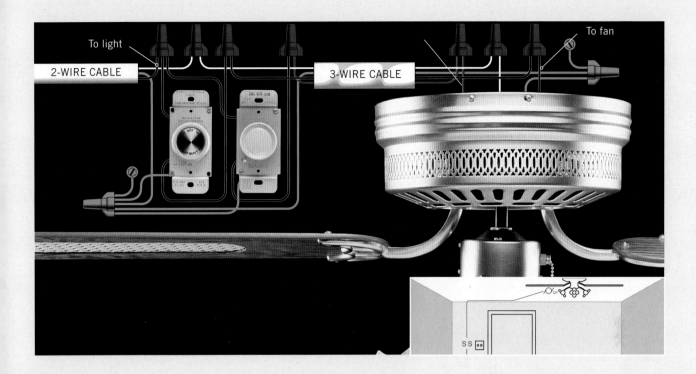

31. Ceiling Fan/Light Fixture Controlled by Ganged Switches
(Switches at End of Cable Run)

Use this switch loop layout variation when it is more practical to install the ganged speed control and dimmer switches for the ceiling fan at the end of the cable run. Requires two-wire and parallel runs of two-wire cables.

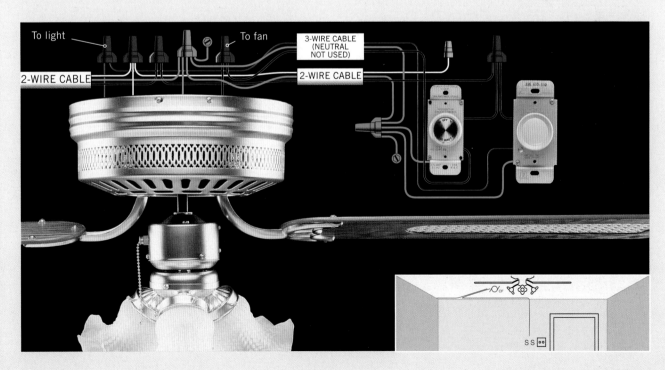

GFCI & AFCI Breakers

Understanding the difference between GFCI (ground-fault circuit interrupter) and AFCI (arc fault circuit interrupter) is tricky for most homeowners. Essentially it comes down to this: arc-fault interrupters keep your house from burning down; ground-fault interrupters keep people from being electrocuted. AFCIs and GFCIs may not be substituted for each other.

The National Electric Code (NEC) requires that an AFCI breaker be installed on most branch circuits that supply outlets or fixtures in newly constructed homes. The NEC also requires adding AFCI protection to these circuits when you add new circuits and modify or extend existing circuits. They're a prudent precaution in any home, especially if it has older wiring. AFCI breakers will not interfere with the operation of GFCI receptacles, so it is safe to install an AFCI breaker on a circuit that contains GFCI receptacles.

Ground-Fault Circuit-Interrupters

A GFCI is an important safety device that disconnects a circuit in the event of a ground fault (when current takes a path other than the neutral back to the panel).

On new construction and when adding or extending electrical circuits, GFCI protection is required for 120-volt and 240-volt receptacles and equipment in these locations: kitchen counter tops and similar work surfaces, dishwasher branch circuit, bathrooms, whirlpool bathtubs, garages, unfinished basements, crawl spaces, lights in crawl spaces, outdoors, within six feet of sinks, and in unfinished accessory buildings such as storage and work sheds and swimming pools, spas, and hot tubs whether indoors or outdoors. In general it is a good practice to protect all receptacle and fixture locations that could encounter damp or wet circumstances.

ARC-FAULT CIRCUIT INTERRUPTERS

AFCIs detect arcing (sparks) that can cause fires between and along damaged wires. AFCI protection is required for 15- and 20-amp, 120-volt circuits that serve living rooms, family rooms, dens, parlors, libraries, dining rooms, bedrooms, sun rooms, kitchens, laundry areas, closets, hallways, and similar rooms and when adding to or extending any of these circuits. AFCI protection is not required for circuits serving bathrooms, garages, the exterior of the home, and appliances such as furnaces and air handlers.

The easiest way to provide AFCI protection for a circuit is to install an AFCI circuit breaker labeled as a "combination" device in the electrical panel. The 2023 NEC allows several alternate methods of providing AFCI protection, but you should consult an electrician before using these alternate methods. You should install combination AFCI circuit breakers when installing new circuits that require AFCI protection. You should install either combination AFCI circuit breakers or AFCI receptacles when you modify, replace, or extend an existing circuit that requires AFCI protection.

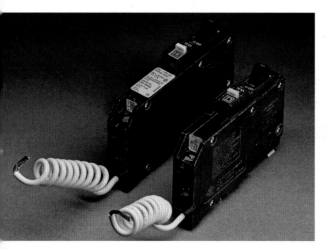

AFCI breakers (left) are similar in appearance to GFCI breakers (right), but they function differently. AFCI breakers trip when they sense an arc fault. GFCI breakers trip when they sense fault between the hot wire and the ground. Combination AFCI/GFCI circuit breakers (center) are also available.

An AFCI-protected receptacle

How to Install an AFCI or GFCI Breaker

1

Locate the breaker for the circuit you'd like to protect. Turn off the main circuit breaker. Remove the cover from the panel, and test to ensure that power is off. Remove the breaker you want to replace from the panel. Remove the black wire from the LOAD terminal of the breaker.

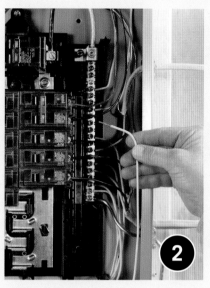

2

Find the white wire on the circuit you want to protect, and remove it from the neutral terminal bar.

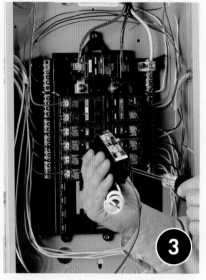

3

Flip the handle of the new AFCI or GFCI breaker to OFF. Loosen both of the breaker's terminal screws. Connect the white circuit wire to the breaker terminal labeled PANEL NEUTRAL. Connect the black circuit wire to the breaker terminal labeled LOAD POWER.

4

Connect the new breaker's coiled white wire to the neutral terminal bar on the service panel.

5

Make sure all the connections are tight. Snap the new breaker into the terminal bar.

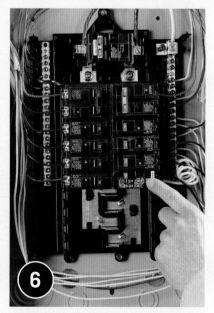

6

Turn the main breaker on. Turn off and unplug all fixtures and appliances on the AFCI or GFCI breaker circuit. Turn the AFCI or GFCI breaker on. Press the test button. If the breaker is wired correctly, the breaker trips open. If it doesn't trip, check all connections or consult an electrician. Replace the panel cover.

Service Panels

Replacing an old 60- or 100-amp electrical service panel with a new 200-amp panel is an ambitious project that requires a lot of forethought. Upgrading your electrical service panel from 100 amps to 200 amps is an ambitious project that requires a lot of forethought. The first step is to obtain a permit. When you are ready to begin, you will need to have your utility company disconnect your house from electrical service at the transformer that feeds your house. When you schedule this, talk to your utility company about the size of your service drop or lateral. That may need to be upgraded too.

Also check with your utility company to make sure you know what equipment is theirs and what belongs to you. In most cases, the electric meter and everything on the street side belongs to the power company, and the meter base and everything on the house side is yours. Be aware that if you tamper with the sealed meter in any way, you likely will be fined. Utility companies will not re-energize your system without approval from your inspecting agency.

Before

Note: The NEC requires outdoor emergency disconnects for home services in new construction, homes undergoing renovation, and all service replacements. An exterior disconnect allows emergency crews to shut off the power. Approved disconnects include
• Service disconnects
• Meter disconnects
• Listed disconnect switches or circuit breakers

Disconnects may be located before or after the meter base and may be in their own enclosure. They must be clearly labeled. Check with the local code authority for specific requirements.

After

Modern homeowners consume more power than our forebears, and it is often necessary to upgrade the electrical service to keep pace. While homeowners are not allowed to make the final electrical service connections, removing the old panel and installing the new panel and meter base yourself can save you hundreds or even thousands of dollars.

TOOLS & MATERIALS

200-amp service panel	Emergency disconnect	Drill/driver
200-amp bypass meter base	Weatherhead	Tape
Circuit breakers	Service cable	Allen wrench
Schedule 80 or RMC conduit and fittings	Circuit wires	Circuit tester
	Plywood backer board	Multimeter
	Screwdrivers	

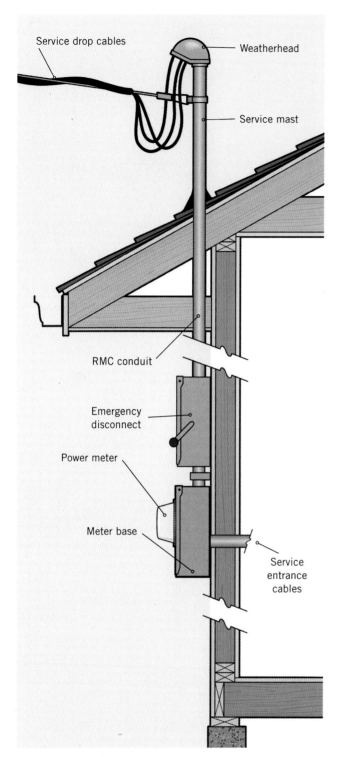

Service drop cables

Weatherhead

Service mast

RMC conduit

Emergency disconnect

Power meter

Meter base

Service entrance cables

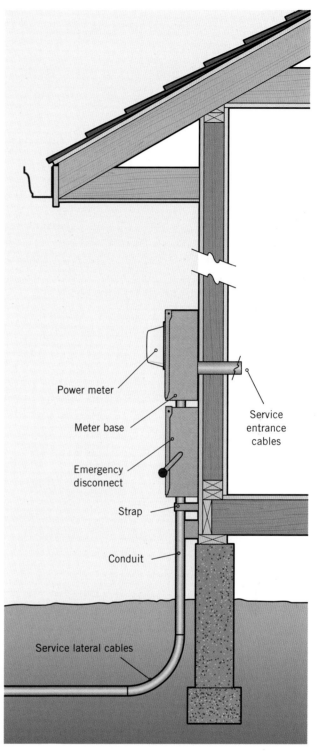

Power meter

Meter base

Emergency disconnect

Strap

Conduit

Service entrance cables

Service lateral cables

Aboveground service drop. In this common configuration, the service cables from the closest transformer (called the service drop) connect to service entrance wires near the weatherhead. This connection is called the service point and is where your property usually begins. The service entrance wires from the weatherhead are routed to a power meter that's owned by your utility company but is housed in a base that's considered your property. From the meter, the service entrance wires enter your house through the wall and are routed to the main service panel, where they are connected to the main circuit breaker.

Underground service lateral. Increasingly, homebuilders are choosing to have power supplied to their new homes underground instead of an overhead service drop. Running the cables in the ground eliminates problems with power outages caused by ice accumulation or fallen trees, but it entails a completely different set of cable and conduit requirements. For the homeowner, however, the differences are minimal, because the hookups are identical once the power service reaches the meter.

Local codes dictate where the main service panel may be placed relative to other parts of your home. Although the codes vary (and always take precedence), national codes stipulate that a service panel (or any other distribution panel) may not be located near flammable materials, in a bathroom, clothes closet or other area designated for storage, above stairway steps, or directly above a workbench or other permanent work station or appliance. The panel also can't be located in a crawl space. If you are installing a new service entry hookup, there are many regulations regarding height of the service drop and the meter. Contact your local inspections office for specific regulations.

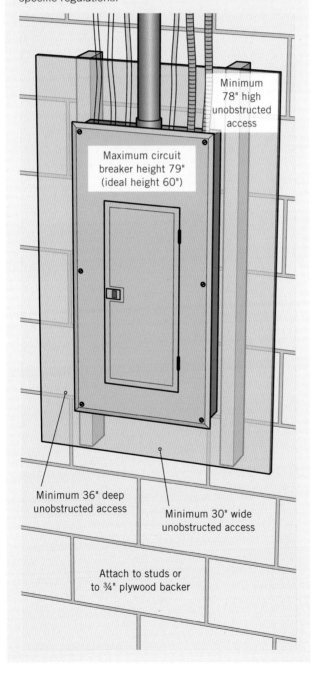

Minimum 78" high unobstructed access

Maximum circuit breaker height 79" (ideal height 60")

Minimum 36" deep unobstructed access

Minimum 30" wide unobstructed access

Attach to studs or to ¾" plywood backer

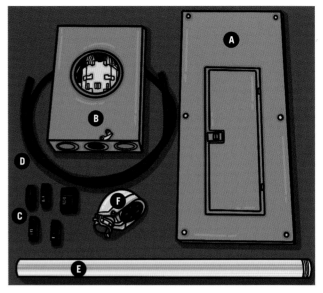

All the equipment you'll need to upgrade your main panel is sold at most larger building centers. It includes (A) a new 200-amp panel; (B) a 200-amp meter base (also called a socket); (C) individual circuit breakers; (D) new, THW, THHW, THWN-2. RHW, RHW-2, XHHW 3/0 copper or 4/0 aluminum; (E) 2" dia. rigid metallic conduit; (F) weatherhead for mast. (E) and (F) are not necessary if you have underground service wires (a service lateral). An exterior service disconnecting means, as shown in the illustration below, will also be necessary.

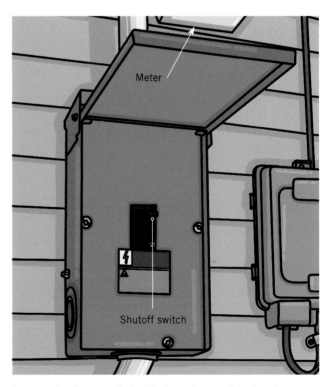

Meter

Shutoff switch

A means to disconnect electrical service must be located outside. This may be the main circuit breaker (service equipment), or it may be a separate switch or circuit breaker. This disconnecting means must be labeled as described in code.

How to Replace a Main Panel

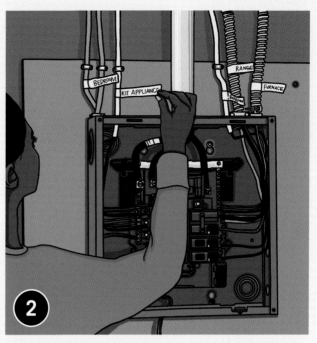

Shut off power to the house at the transformer. This must be done by a technician who is certified by your utility company. Also have the utility worker remove the old meter from the base. It is against the law for a homeowner to break the seal on the meter.

Label all incoming circuit wires before disconnecting them. Labels should be written clearly on tape that is attached to the cables outside of the existing panel. Test the circuits before starting to make sure they are labeled correctly.

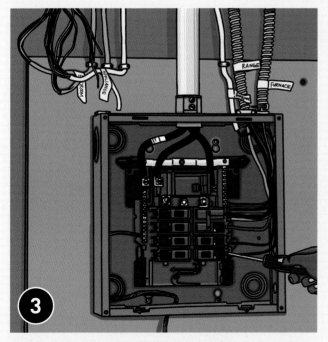

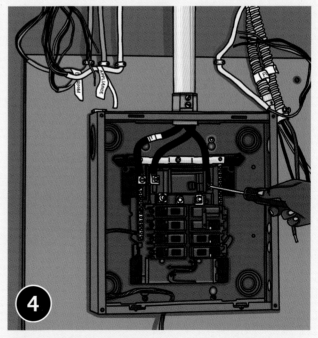

Disconnect incoming circuit wires from breakers, grounding bar, and neutral terminal bar. Also disconnect cable clamps at the knockouts on the panel box. Retract all circuit wires from the panel and coil it up neatly, with the labels clearly visible.

Unscrew the lugs securing the service entry cables at the top of the panel. For 240-volt service you will find two heavy-gauge SE cables, probably with black sheathing. Each cable carries 120 volts of electricity. A neutral service cable, usually of smaller gauge than the SE cables, will be attached to the neutral terminal bar. This cable returns current to the source.

(continued)

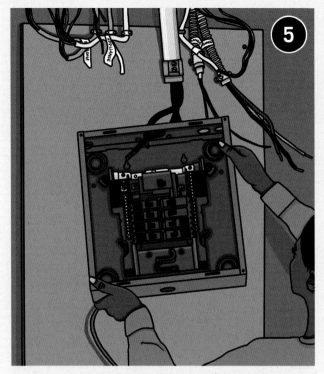

Remove the old service panel box. Boxes are rated for a maximum current capacity; and if you are upgrading, the components in the old box will be undersized for the new service levels. The new box will have a greater number of circuit slots as well.

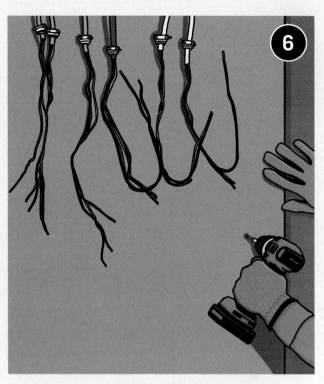

Replace the old panel backer board with a larger board in the installation area. A piece of ¾" plywood is typical. Make sure the board is well secured at wall framing members.

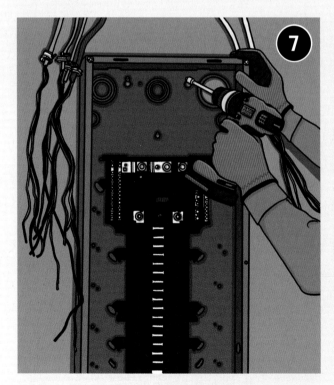

Attach the subpanel box to the backer board, making sure that at least two screws are driven through the backer and into wall studs. Drill clearance holes in the back of the box at stud locations if necessary. Use roundhead screws that do not have tapered shanks so the screwhead seats flat against the panel.

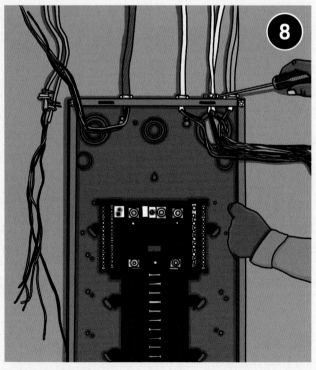

Attach properly sized cable clamps to the box at the knockout holes. Install one cable per knockout in this type of installation and plan carefully to avoid removing knockouts that you do not need to remove (if you do make a mistake, you can fill the knockout hole with a plug).

SPLICING IN THE BOX

Some wiring codes allow you to make splices inside the panel box if the circuit wire is too short. Use the correct wire cap and wind electrical tape over the conductors where they enter the cap. If your municipality does not allow splices in the panel box, you'll have to rectify a short cable by splicing it in a junction box before it reaches the panel and then replacing the cable with a longer section for the end of the run. Make sure each circuit line has at least 12" of slack.

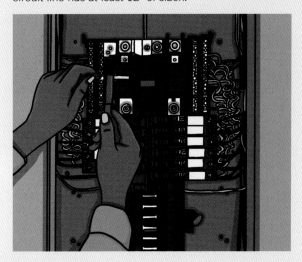

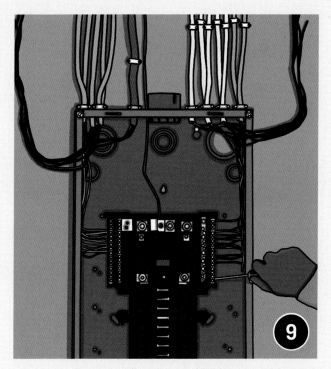

Attach the white neutral from each circuit cable to the neutral terminal bar. Most panels have a preinstalled neutral terminal bar, but in some cases you may need to purchase the bar separately and attach it to the panel back. The panel should also have a separate grounding bar that you also may need to purchase separately.

Note: For GFCI and AFCI breakers, the neutral circuit wire connects to the breaker, and the breaker's coiled neutral lead connects to the neutral terminal bar.

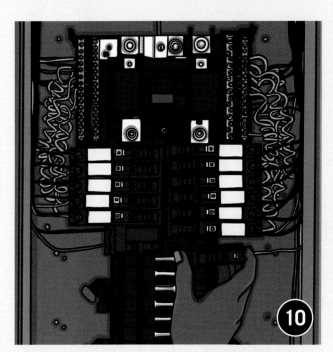

Attach the hot lead wire to the terminal on the circuit breaker, and then snap the breaker into an empty slot. When loading slots, start at the top of the panel and work your way downward. It is important that you balance the circuits as you go to equalize the amperage. For example, do not install all the 15-amp circuits on one side and all the 20-amp circuits on the other.

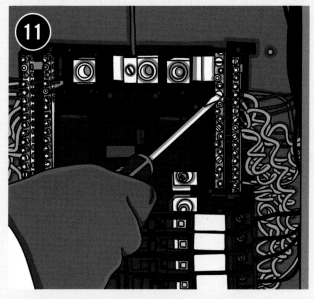

Install grounding conductors. Local codes are very specific about how the grounding and bonding needs to be accomplished. For example, some require multiple rods driven at least 6 ft. apart. Discuss your grounding requirements thoroughly with your inspector or an electrician before making your plan.

(continued)

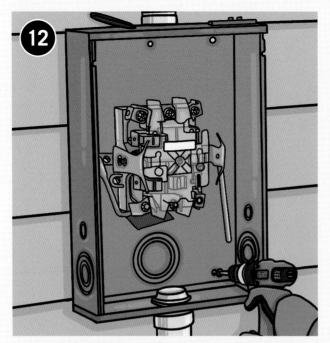

Remove the old meter base and install the new 200-amp meter base. A meter base is usually installed near eye-level.

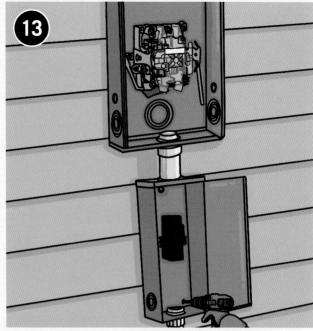

Install the new emergency disconnect. Install it above the meter base for overhead service or below the meter base for underground service (see page 51). The switch or breaker should not be more than 79 inches above the ground. Install the service mast conduit (overhead service) or the service lateral conduit (underground service). Install the conduit between the meter base and the emergency disconnect and between the meter base and your house.

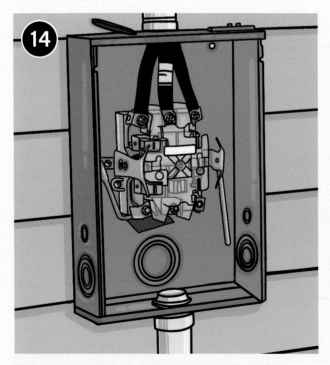

If you have an overhead service drop, install the service entrance wires from the overhead service drop. Install the feeder wires between the meter base and your new panel. Code is very specific about how these connections are made. In most cases, you'll need to tighten the terminal nuts with a specific amount of torque that requires a torque wrench to measure.

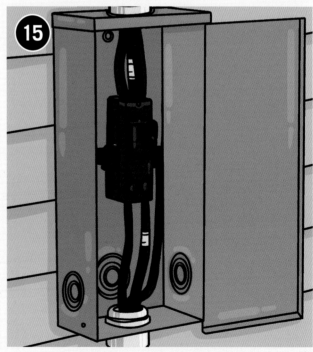

If you have an underground service lateral, install the service entrance wires from the underground service lateral. Install the feeder wires between the meter base and your new panel.

16

Install the wires between the meter base and the emergency disconnect.

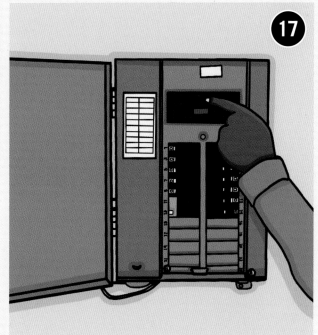

17

Connect the feeder wires to your new panel, then have the panel and all connections inspected and approved by your local building department. After you pass all inspections, contact the public utility company to make the connections at the power transformer.

 TALL MAST, SHORT ROOF

The service drop must occur at least 10 ft. above ground level, and as much as 14 ft. in some cases. Occasionally, this means that you must run the conduit for the service mast up through the eave of your roof and seal the roof penetration with a boot.

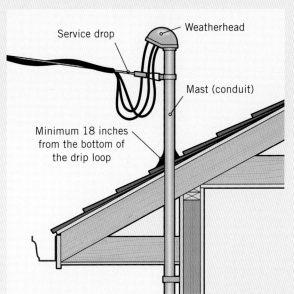

Service drop — Weatherhead

Mast (conduit)

Minimum 18 inches from the bottom of the drip loop

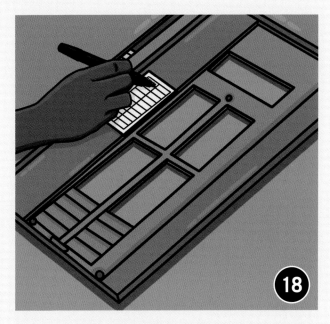

18

Create an accurate circuit index and affix it to the inside of the service panel door. List all loads that are on the circuit as well as the amperage. Once you have restored power to the new panel, test out each circuit to make sure you don't have any surprises. With the main breakers on, shut off all individual circuit breakers, and then flip each one on by itself. Walk through your house and test every switch and receptacle to confirm the loads on that circuit.

Grounding & Bonding a Wiring System

All home electrical systems must be bonded and grounded according to code standards. This entails two tasks: the metal water and gas pipes must be connected electrically to create a continuous low resistance path back to the main electrical panel; and the main electrical panel must be grounded to a grounding electrode such as a ground rod or rods driven into the earth near the foundation of your house. Although the piping system is bonded to the ground through your main electrical service panel, the panel grounding and the piping bonding are unrelated when it comes to function. The grounding wire that runs from your electrical panel to grounding electrode helps even out voltage increases that often occur because of lightning and other causes. The wires that bond your metal piping are preventative, and they only become important in the unlikely event that an electrical conductor energizes the pipe. In that case, correct bonding of the piping system will ensure that the current does not remain in the system, where it could shock anyone who touches a part of the system, such as a faucet handle. Bonding is done relatively efficiently at the water heater, as the gas piping and water piping are typically located there.

Gas pipe in older homes is usually steel or copper. The bonding connection point for these pipes can be at any accessible location, such as at the water heater or at the gas meter. Gas pipe in some new homes is a flexible material called corrugated stainless-steel tubing (CSST). Refer to the manufacturer's instructions for bonding the CSST.

A pair of 8-ft.-long metal ground rods are driven into the earth next to your house to provide a path to ground for your home wiring system.

TOOLS & MATERIALS

Hammer	½" drill bit	3 pipe ground clamps	Grounding rods
Flat screwdriver	A length of ground wire	Eye and ear protection	5-lb. maul
Drill	Wire staples	Work gloves	Caulk

How to Bond Metallic Piping

Determine the amperage rating of your electrical service by looking at your main breakers. (If you have an older system and are unsure about its amperage rating, consult an electrician.) The system amperage (usually 100 or 200 amps) determines the required gauge of the bonding wire you need. #6 copper wire is often sufficient for service not exceeding 200 amps. Always confirm the correct gauge with the local electrical inspector.

Run the bonding wire from a point near your water heater to an exit point where the wire can be bonded to the grounding wire that leads to the exterior grounding electrodes. This is frequently done at the service panel. Run this wire as you would any other cable, leaving approximately 6 to 8 ft. of wire at the water heater. If you are running this wire through the ceiling joists, drill a ½" hole as close to the center as possible to not weaken the joist. Secure the wire every 2 ft. if running it parallel to the joists.

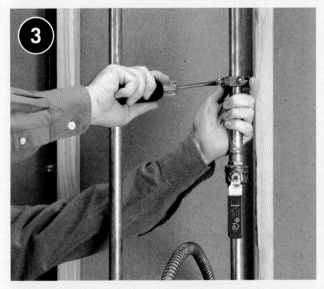

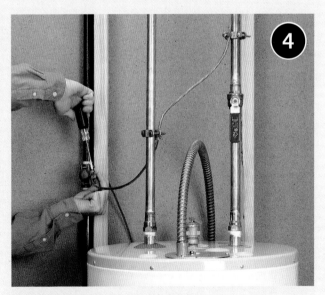

Install pipe ground clamps on each pipe (hot water supply, cold water supply, gas), roughly 1 foot above the water heater. Do not install clamps near a union or elbow because the tightening of the clamps could break or weaken soldered joints. Also make sure the pipes are free and clear of any paint, rust, or any other contaminant that may inhibit a good clean connection. Do not overtighten the clamps. Use clamps that are compatible with the pipe so that corrosion will not occur. Use copper or brass clamps on copper pipe. Use brass or steel clamps on steel pipe.

Route the ground wire through each clamp wire hole and then tighten the clamps onto the wire. Do not cut or splice the wire: the same wire should run through all clamps.

(continued)

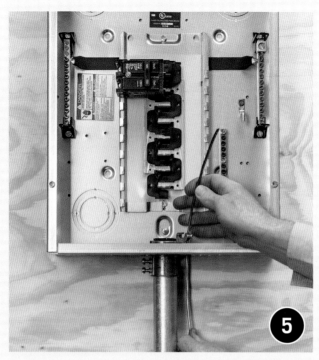

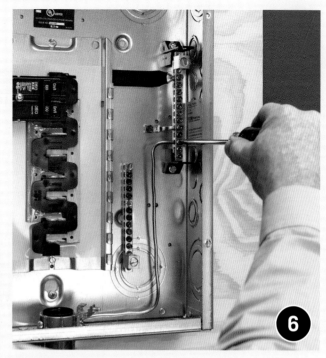

At the panel, turn off the main breaker. Open the cover by removing the screws, and set the cover aside. Route the ground wire through a small ⅜" hole provided toward the rear of the panel on the top or bottom. You will usually have to knock the plug out of this hole by placing a screwdriver on it from the outside and tapping with a hammer. Make sure the ground wire will not come into contact with the terminal bars in the middle of the panel or any of the load terminals on the breakers.

Locate an open hole on your ground and neutral terminal and insert the ground wire. These holes are large enough to accommodate up to a #4 awg wire, but it may be difficult at times. If you're having trouble pushing the wire in, trim a little wire off the end and try with a clean cut piece. Secure the set screw at the lug. Replace the panel cover and turn the main breaker back on.

Tips for Grounding Service Panels

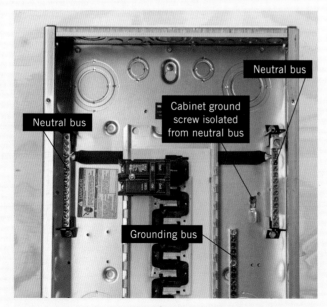

Neutral bus

Cabinet ground screw isolated from neutral bus

Neutral bus

Grounding bus

In a subpanel, the grounding terminal should be bonded to the subpanel cabinet. The neutral terminal should not be bonded to the subpanel cabinet.

Metallic conduit must be physically and electrically connected to panel cabinets. A bonding bushing may be required in some cases, where not all of a knockout is removed.

Ground Rod Installation

The ground rod is an essential part of the grounding system. Its primary function is to create a path to ground for electrical current, such as lightning, line surges, and unintentional contact with high voltage lines. If you upgrade your electrical service, you likely will need to upgrade your grounding wire and rods to meet current code.

Call before you dig! Make sure the area where you will be installing the ground rods is free and clear from any underground utilities.

NOTE: Different municipalities have different requirements for grounding, so be sure to check with the AHJ (Authority Having Jurisdiction) first before attempting to do this yourself.

TOOLS & MATERIALS

⅝" × 8' ground rods	5-pound maul	Screwdriver	Wire cutters
Drill ¼	Copper ground wire (size as required by local code)	(2) brass (acorn) clamp	Caulk
5⁄16" drill bit		Pliers	
Ladder			

How to Install a Grounding Electrode System

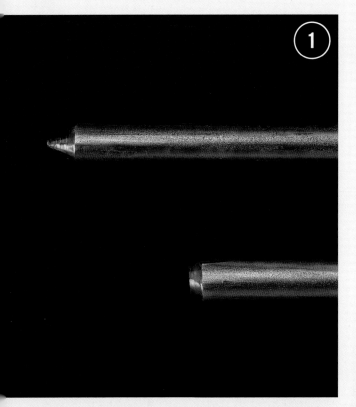

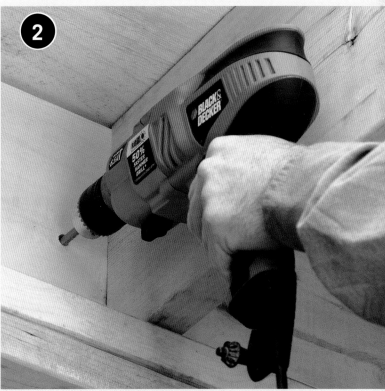

Begin by purchasing two copper-coated steel ground rods ⅝" diameter by 8' long. Grounding rods have a driving point on one end and a striking face on the other end.

Drill a 5⁄16" hole in the rim joist of your house, as close as practical to the main service panel to the outside of the house above the ground level at least 6". *(continued)*

About 1 foot from the foundation of the house, pound one ground rod into the earth with a 5-lb. maul. If you encounter a rock or other obstruction, you can pound the ground rod at an angle as long as it does not exceed 45°. Drive until only 3" or 4" of the rod is above ground. Measure at least 6 ft. from the first ground rod and pound in another one.

Run uninsulated copper ground wire from the ground bus in your main service panel through the hole in the rim joist and to the exterior of the house, leaving enough wire to connect the two ground rods together.

Using a brass clamp commonly referred to as an acorn, connect the wire to the first ground rod, pulling the wire taut so no slack exists. Continue pulling the wire to reach the second grounding rod, creating a continuous connection.

Connect the second ground rod with another acorn to the uncut grounding wire previously pulled through the first acorn. Trim the excess wire.

7

Dig out a few inches around each rod to create clearance for the 5-lb. maul. Creating a shallow trench beneath the grounding wire between the rods is also a good idea. Drive each rod with the maul until the top of the rod is a few inches below grade.

Inject caulk into the hole in the rim joist on both the interior and exterior side.

Tips for Grounding & Bonding

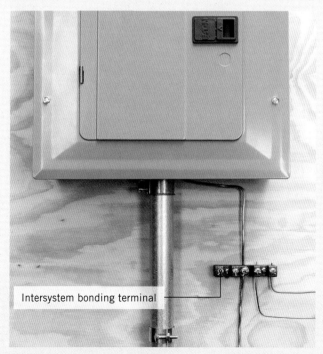

Intersystem bonding terminal

Use an intersystem bonding terminal to ground nonelectrical systems such as telephone and cable.

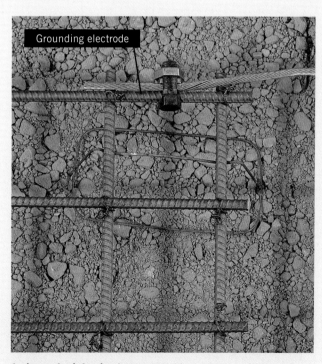

Grounding electrode

A piece of reinforcing bar encased in a concrete footing is a common grounding electrode in new construction. Called an ufer, the electrode must be No. 4 or larger rebar and at least 20 ft. long. (Shown prior to pouring concrete.)

Subpanels

Single-pole circuit breakers

Before

Install circuit breaker subpanels if the main circuit breaker panel does not have enough open breaker slots for the new circuits you are planning. Subpanels serve as additional distribution centers for connecting circuits. They receive power from a double-pole circuit breaker you install in the main circuit breaker panel.

If the main service panel is so full that there is no room for the double-pole subpanel breaker, you can reconnect some of the existing 120-volt circuits to special slimline breakers (see photo at right). You may be required to install AFCI breakers for the new circuits. AFCI breakers are full size breakers. Be sure to plan for this when estimating the space left in your existing main service panel and all subpanels.

Plan your subpanel installation carefully, making sure your electrical service supplies enough power to support the extra load of the new subpanel circuits. Assuming your main service is adequate, consider installing a subpanel that's a little larger than you need to provide enough extra amps to meet the needs of future wiring projects.

Also consider the physical size and the current rating of the subpanel, and choose one that has enough extra slots and current capacity to hold circuits you may want to install later. The smallest panels have room for up to six single-pole breakers (or three double-pole breakers), while the largest models can hold 20 single-pole breakers or more.

Subpanels often are mounted near the main circuit breaker panel. Or, for convenience, they can be installed close to the areas they serve, such as in a new room addition. In a finished room, a subpanel can be painted or housed in a decorative cabinet so it is less of a visual distraction—just make sure it's accessible.

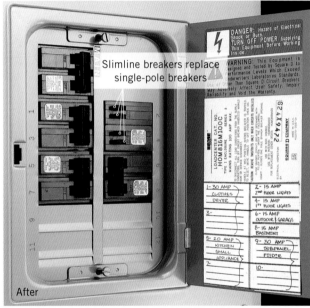

Slimline breakers replace single-pole breakers

After

To conserve space in a service panel, you may be able to replace existing single-pole breakers with slimline breakers. Slimline breakers take up half the space of standard breakers, allowing you to fit two circuits into one single slot on the service panel. In the service panel shown above, four single-pole 120-volt breakers were replaced with slimline breakers to provide the double opening needed for a 30-amp, 240-volt subpanel feeder breaker. Use slimline breakers (if your municipality allows them) with the same amp rating as the standard single-pole breakers you are removing, and make sure they are approved for use in your panel. If your municipality and panel allow slimline breakers, there may be restrictions on the quantity and location where they may be installed on the panel.

TOOLS & MATERIALS

Hammer	Cable ripper	Cable clamps	Double-pole circuit breaker
Screwdriver	Combination tool	Three-wire NM cable	Circuit breaker subpanel
Voltage tester	Screws	Cable staples	Slimline circuit breakers

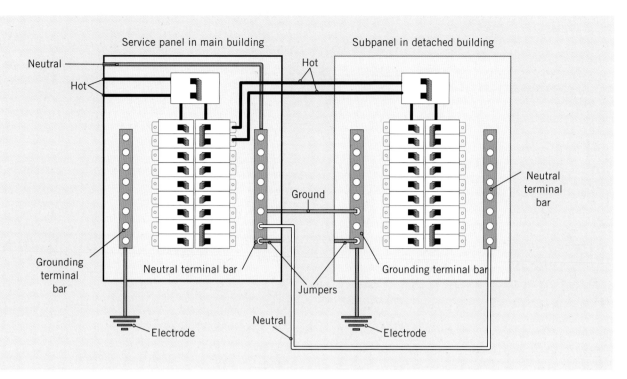

Service panel in main building — Subpanel in detached building

Neutral — Hot — Hot — Ground — Neutral terminal bar — Grounding terminal bar — Neutral terminal bar — Neutral terminal bar — Grounding terminal bar — Jumpers — Electrode — Neutral — Electrode

Wiring diagram for wiring a feeder from the main service panel to a subpanel in a separate building.

 ## How to Install a Subpanel

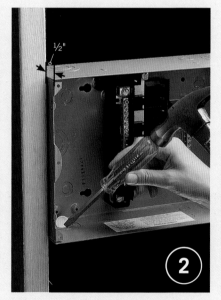

Subpanels are subject to the same installation and clearance rules as service panels. The subpanel can be mounted to the sides of studs or to plywood attached between two studs. The panel shown here extends ½" past the face of studs so it will be flush with the finished wall surface. Follow the manufacturer's installation specifications.

Open a knockout in the subpanel using a screwdriver and hammer. Run the feeder cable from the main circuit breaker panel to the subpanel, leaving about 2 ft. of excess cable at each end.

Attach a cable clamp to the knockout in the subpanel. Insert the cable into the subpanel, and then anchor it to framing members within 8" of each panel and every 54" thereafter.

(continued)

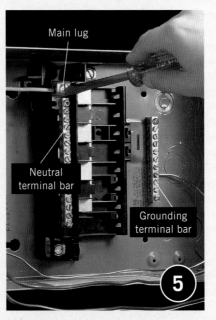

Strip away outer sheathing from the feeder cable using a cable ripper. Leave at least ¼" of sheathing extending into the subpanel. Tighten the cable clamp screws so the cable is held securely, but not so tightly that the wire sheathing is crushed.

Strip ½" of insulation from the white neutral feeder wire, and attach it to the main lug on the subpanel neutral terminal bar. Connect the grounding wire to a setscrew terminal on the grounding terminal bar. Fold excess wire around the inside edge of the subpanel.

Strip away ½" of insulation from the red and the black feeder wires. Attach one wire to the main lug on each of the hot terminal bars. Fold excess wire around the inside edge of the subpanel.

At the main circuit breaker panel, shut off the main circuit breaker, and then remove the coverplate and test for power. If necessary, make room for the double-pole feeder breaker by removing single-pole breakers and reconnecting the wires to slimline circuit breakers. Open a knockout for the feeder cable using a hammer and screwdriver.

NOTE: Some panels do not allow slimline breakers and some restrict where slimline breakers can be installed. Read the instructions on the panel cover.

Strip away the outer sheathing from the feeder cable so that at least ¼" of sheathing will reach into the main service panel. Attach a cable clamp to the cable, and then insert the cable into the knockout, and anchor it by threading a locknut onto the clamp. Tighten the locknut by driving a screwdriver against the lugs. Tighten the clamp screws so the cable is held securely, but not so tightly that the cable sheathing is crushed.

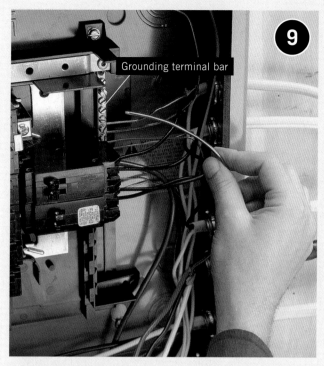

⑨

Grounding terminal bar

Bend the bare copper wire from the feeder cable around the inside edge of the main circuit breaker panel, and connect it to one of the setscrew terminals on the grounding terminal bar.

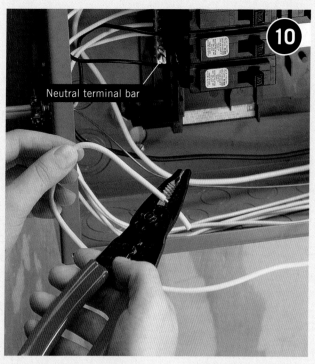

⑩

Neutral terminal bar

Strip away ½" of insulation from the white feeder wire. Attach the wire to one of the setscrew terminals on the neutral terminal bar. Fold excess wire around the inside edge of the service panel.

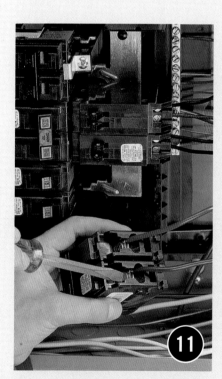

⑪

Strip ½" of insulation from the red and the black feeder wires. Attach one wire to each of the setscrew terminals on the double-pole feeder breaker.

NOTE: If your subpanel arrived with a preinstalled grounding screw in the panel back, remove and discard it.

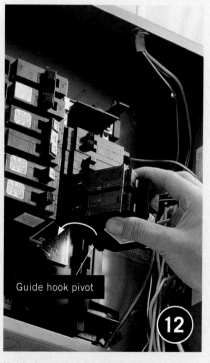

Guide hook pivot

⑫

Hook the end of the feeder circuit breaker over the guide hooks on the panel, and then push the other end forward until the breaker snaps onto the hot terminal bars (follow manufacturer's directions). Fold excess wire around the inside edge of the circuit breaker panel.

⑬

If necessary, remove two tabs from the cover plate where the double-pole feeder breaker will fit, and then reattach the cover plate. Label the feeder breaker on the circuit index. Turn the main breaker on, but leave the feeder breaker off until all subpanel circuits have been connected and inspected.

Underfloor Radiant Heat Systems

Floor-warming systems require very little energy to run and are designed to heat ceramic tile floors only; they generally are not used as sole heat sources for rooms. Extending a branch circuit or adding a new branch to install new receptacles, lights, switches, or equipment requires a permit. Check with the electrical inspector before starting such projects.

A typical floor-warming system consists of one or more thin mats containing electric resistance wires that heat up when energized, like an electric blanket. The mats are installed beneath the tile and are hardwired to a 120-volt GFCI circuit. A thermostat controls the temperature, and a timer turns the system off automatically.

The system shown in this project includes two plastic mesh mats, each with its own power lead that is wired directly to the thermostat. Radiant mats may be installed over a plywood subfloor, but if you plan to install floor tile, you should put down a base of cementboard first, and then install the mats on top of the cementboard.

A crucial part of installing this system is to use a multimeter to perform several resistance checks to make sure the heating wires have not been damaged during shipping or installation.

Electrical service required for a floor-warming system is based on size. A smaller system may connect to an existing circuit, but this may not be a bathroom receptacle circuit, and the system may not draw more than 50 percent of the circuit current capacity. A larger system will need a dedicated circuit; follow the manufacturer's instructions. These systems should be on a GFCI-protected circuit.

To order a floor-warming system, contact the manufacturer or dealer (see Resources, page 125). In most cases, you can send them plans and they'll custom-fit a system for your project area.

TOOLS & MATERIALS

Vacuum cleaner

Multimeter

Tape measure

Scissors

Router/rotary tool

Marker

Electric wire fault indicator (optional)

Hot glue gun

Radiant floor mats

12/2 NM cable

Trowel or rubber float

Conduit

Thinset mortar

Thermostat with sensor

Junction box(es)

Tile or stone floorcovering

Drill

Double-sided carpet tape

Cable clamps

A radiant floor-warming system employs electric heating mats that are covered with floor tile to create a floor that's cozy underfoot.

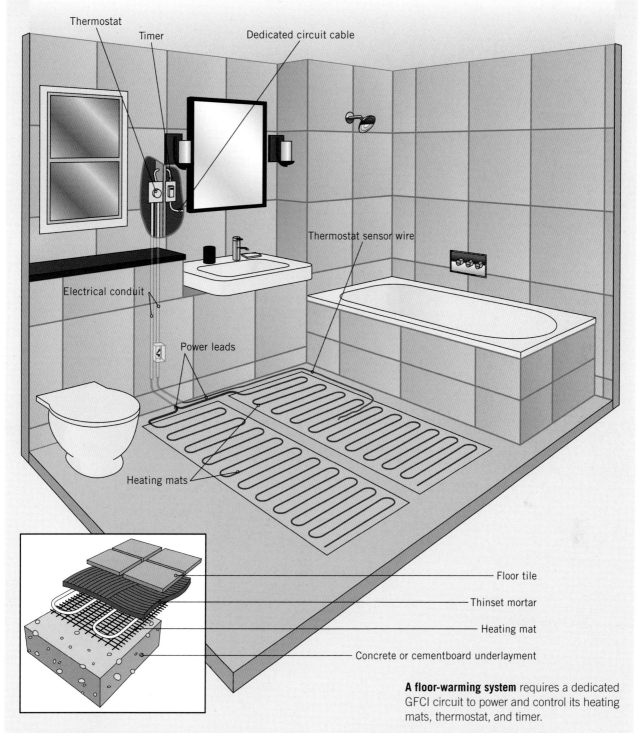

Thermostat

Timer

Dedicated circuit cable

Thermostat sensor wire

Electrical conduit

Power leads

Heating mats

Floor tile

Thinset mortar

Heating mat

Concrete or cementboard underlayment

A floor-warming system requires a dedicated GFCI circuit to power and control its heating mats, thermostat, and timer.

- Each radiant mat must have a direct connection to the power lead from the thermostat, with the connection made in a junction box in the wall cavity. Do not install mats in series.

- Do not install radiant floor mats under shower areas.

- Do not overlap mats or let them touch.

- Do not cut heating wire or damage heating wire insulation.

- The distance between wires in adjoining mats should equal the distance between wire loops measured center to center.

Installing a Radiant Floor-Warming System

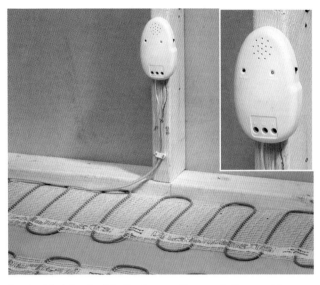

Floor-warming systems must be installed on a circuit with adequate amperage and a GFCI breaker. Smaller systems may tie into an existing circuit, but larger ones need a dedicated circuit. Follow local building and electrical codes that apply to your project.

An electric wire fault indicator monitors each floor mat for continuity during the installation process. If there is a break in continuity (for example, if a wire is cut), an alarm sounds. If you choose not to use an indicator tool to monitor the mat, test for continuity frequently using a multimeter.

How to Install a Radiant Floor-Warming System

Install electrical boxes to house the thermostat and timer. In most cases, the box should be located 60" above floor level. Use a 4"-deep × 4"-wide double-gang box for the thermostat/timer control if your kit has an integral model. If your timer and thermostat are separate, install a separate single box for the timer.

Drill access holes in the sole plate for the power leads that are preattached to the mats (they should be over 10 ft. long). The leads should be connected to a supply wire from the thermostat in a junction box located in a wall near the floor and below the thermostat box. The access hole for each mat should be located directly beneath the knockout for that cable in the thermostat box. Drill through the sill plate vertically and horizontally so the holes meet in an L-shape.

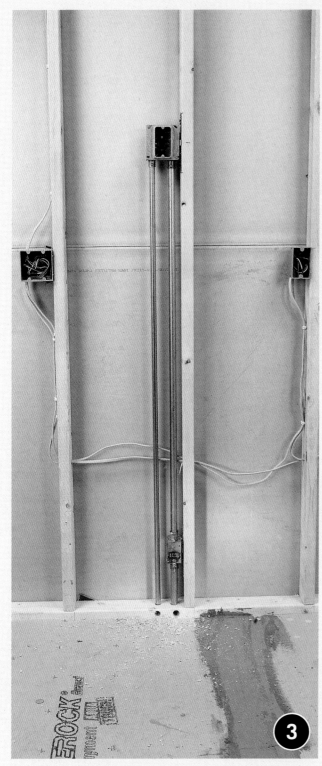

Run conduit from the electrical boxes to the sill plate. The line for the supply cable should be ¾" conduit. If you are installing multiple mats, the supply conduit should feed into a junction box about 6" above the sill plate and then continue into the ¾" hole you drilled for the supply leads. The sensor wire needs only ½" conduit that runs straight from the thermostat box via the thermostat. Unless you are tapping into an existing circuit, the mats should be powered by a dedicated 20-amp GFCI circuit of 12/2 NM cable run from your main service panel to the electrical box (this is for 120-volt mats—check your instruction manual for specific circuit recommendations).

Clean the floor surface thoroughly to get rid of any debris that could potentially damage the wire mats. A vacuum cleaner generally does a more effective job than a broom.

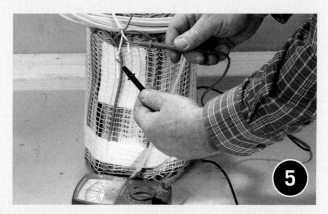

Test for resistance using a multimeter set to measure ohms. This is a test you should make frequently during the installation, along with checking for continuity. If the resistance is off by more than 10% from the theoretical resistance listing (see manufacturer's chart in installation instructions), contact a technical support operator for the kit manufacturer. For example, the theoretical resistance for the 1×50 ft. mat seen here is 19, so the ohms reading should be between 17 and 21.

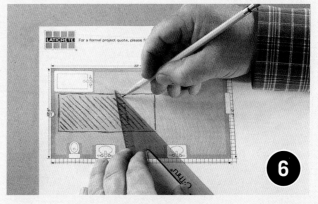

Finalize your mat layout plan. Most radiant floor warming mat manufacturers will provide a layout plan for you at the time of purchase, or they will give you access to an online design tool so you can come up with your own plan. This is an important step to the success of your project, and the assistance is free. *(continued)*

Unroll the radiant mat or mats and allow them to settle. Arrange the mat or mats according to the plan you created. It's okay to cut the plastic mesh so you can make curves or switchbacks, but do not cut the heating wire under any circumstances, even to shorten it.

Finalize the mat layout, and then test the resistance again using a multimeter. Also check for continuity in several different spots. If there is a problem with any of the mats, you should identify it and correct it before proceeding with the mortar installation.

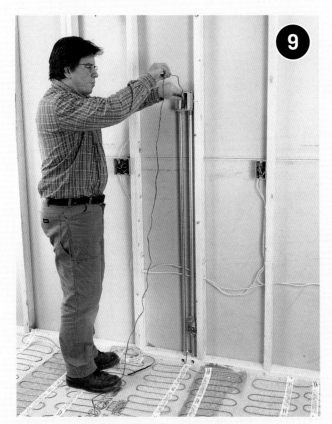

Run the thermostat sensor wire from the electrical box down the ½" conduit raceway and out the access hole in the sill plate. Select the best location for the thermostat sensor, and mark the location onto the flooring. Also mark the locations of the wires that connect to and lead from the sensor.

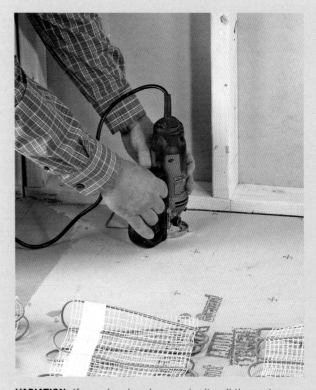

VARIATION: If your local codes require it, roll the mats out of the way, and cut a channel for the sensor and the sensor wires into the floor or floor underlayment. For most floor materials, a spiral cutting tool does a quick and neat job of this task. Remove any debris.

Bond the mats to the floor. If the mats in your system have adhesive strips, peel off the adhesive backing and roll out the mats in the correct position, pressing them against the floor to set the adhesive. If your mats have no adhesive, bind them with strips of double-sided carpet tape. The thermostat sensor and the power supply leads should be attached with hot glue (inset photo) and run up into their respective holes in the sill plate if you have not done this already. Test all mats for resistance and continuity.

Cover the floor installation areas with a layer of thinset mortar that is thick enough to fully encapsulate all the wires and mats (usually around ¼" in thickness). Check the wires for continuity and resistance regularly, and stop working immediately if there is a drop in resistance or a failure of continuity. Allow the mortar to dry overnight.

Connect the power supply leads from the mat or mats to the NM cable coming from the thermostat inside the junction box near the sill. Power must be turned off. The power leads should be cut so about 8" of wire feeds into the box. Be sure to use cable clamps to protect the wires.

Connect the sensor wire and the power supply lead (from the junction box) to the thermostat/timer according to the manufacturer's directions. Attach the device to the electrical box, restore power, and test the system to make sure it works. Once you are convinced that it is operating properly, install floor tiles and repair the wall surfaces.

Note: Enclose the junction box with a blank cover. Do not cover it with drywall; it must remain accessible.

Backup Power Supply

Installing a backup generator is an invaluable way to prepare your family for emergencies. The simplest backup power system is a portable gas-powered generator and an extension cord or two. A big benefit of this approach is that you can run a refrigerator and a few worklights during a power outage with a generator that can also be transported to remote job sites or on camping trips when it's not doing emergency backup duty. This is also the least expensive way to provide some backup power for your home. You can purchase a generator at most home centers and be up and running in a matter of hours. If you take this approach, it is critically important that you make certain any loads being run by your generator are disconnected from the utility power source.

The next step up is to incorporate a manual transfer switch for your portable generator. Transfer switches are permanently hardwired to your service panel. They are mounted on either the interior or the exterior of your house between the generator and the service panel. You provide a power feed from the generator into the switch. The switch is wired to selected essential circuits in your house, allowing you to power lights, furnace blowers, and other loads that can't easily be run with an extension cord. But perhaps the most important job a transfer switch performs is to disconnect the utility power. If the inactive utility power line is attached to the service panel, "backfeed" of power from your generator to the utility line can occur when the generator kicks in. This condition could be fatal to line workers who are trying to restore power. The potential for backfeed is the main reason many municipalities insist that only a licensed electrician hook up a transfer switch. Most also require a permit. Using a transfer switch not installed by a professional may also void the warranty of the switch and the generator.

Automatic transfer switches turn on the generator and switch off the utility supply when they detect a significant drop in line voltage. They may be installed with portable generators, provided the generator is equipped with an electric starter.

Large standby generators that resemble central air conditioners are the top of the line in backup power supply systems. Often fueled by home natural gas lines that offer a bottomless fuel source or in-yard propane tanks, standby generators are made in sizes with as much as 20 to 40 kilowatts of output—enough to supply all of the power needs of a 5,000-square-foot home.

NEC requirements for generators include that the generator receptacles should be GFCI protected. The generator should be equipped with a means to shut it down in an emergency and render it incapable of restarting without a manual reset.

Generators have a range of uses. Large hardwired models can provide instant emergency power for a whole house. Smaller models (below) are convenient for occasional short-term backup as well as job sites or camping trips.

Choosing a Backup Generator

A 2,000- to 5,000-watt gas-powered generator and a few extension cords can power lamps and an appliance or two during shorter-term power outages. Appliances must not be connected to household wiring and the generator simultaneously. **Never plug a generator into an outlet. Never operate a generator indoors.** Run extension cords through a garage door.

A permanent transfer switch patches electricity from a large portable generator through to selected household circuits via an inlet at your service panel (inset), allowing you to power hardwired fixtures and appliances with the generator.

For full, on-demand backup service, install a large standby generator wired to an automatic transfer panel. In the event of a power outage, the household system instantly switches to the generator.

A Typical Backup System

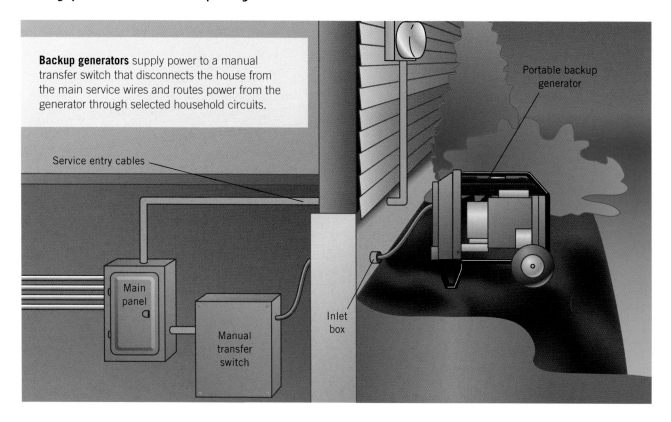

Backup generators supply power to a manual transfer switch that disconnects the house from the main service wires and routes power from the generator through selected household circuits.

Portable backup generator

Service entry cables

Main panel

Manual transfer switch

Inlet box

Choosing a Generator

Choosing a generator for your home's needs requires a few calculations. The chart below gives an estimate of the size of generator typically recommended for a house of a certain size. You can get a more accurate number by adding up the power consumption (the watts) of all the circuits or devices to be powered by a generator. It's also important to keep in mind that, for most electrical appliances, the amount of power required at the moment you flip the ON switch is greater than the number of watts required to keep the device running. For instance, though an air conditioner may run on 5,400 watts of power, it will require a surge of 7,200 watts at startup (the power range required to operate an appliance is usually listed somewhere on the device itself). These two numbers are called run watts (or Rated Load Amps, RLA) and surge watts (or Locked Rotor Amps, LRA). Generators are typically sold according to run watts (a 5,000-watt generator can sustain 5,000 watts). They are also rated for a certain number of surge watts (a 5,000-watt generator may be able to produce a surge of 10,000 watts). If the surge watts aren't listed, ask, or check the manual. Some generators can't develop many more surge watts than run watts; others can produce twice as much surge as run wattage.

It's not necessary to buy a generator large enough to match the surge potential of all your circuits (you won't be turning everything on simultaneously), but surge watts should factor in your purchasing decision. If you will be operating the generator at or near capacity, it is also a wise practice to stagger startups for appliances.

You will need a large amount of gasoline to power a gasoline generator for more than a day or so. Gasoline goes bad over time, so you will need to stock up on gas before a long outage. Be sure to store gasoline well away from any living space. Portable generators powered by propane are available, and may be a better choice for some. Propane can last in a tank for years.

SIZE OF HOUSE (IN SQUARE FEET)	RECOMMENDED GENERATOR SIZE (IN KILOWATTS)
Up to 2,700	5–11
2,701–3,700	14–16
3,701–4,700	20
4,701–7,000	42–47

Types of Transfer Switches

When using a cord-connected switch, consider mounting an inlet box to the exterior wall. This will allow you to connect a generator without running a cord into the house.

Cord-connected transfer switches (shown above) are hard-wired to the service panel (in some cases they're installed after the service panel and operate only selected circuits). These switches contain a male receptacle for a power supply cord connected to the generator. Automatic transfer switches (not shown) detect voltage drop-off in the main power line and switch over to the emergency power source.

GENERATOR TIPS

If you'll need to run sensitive electronics such as computers or home theater equipment, look for a generator with power inverter technology that dispenses "clean power" with a stable sine wave pattern.

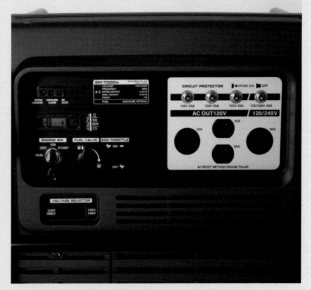

A generator that will output 240-volt service is required to run most central air conditioners. If your generator has variable output (120/240), make sure the switch is set to the correct output voltage.

Running & Maintaining a Backup System

Even with a fully automatic standby generator system fueled by natural gas or propane, you will need to conduct some regular maintenance and testing to make sure all systems are ready in the event of power loss. If you're depending on a portable generator and extension cords or a standby generator with a manual transfer switch, you'll also need to know the correct sequence of steps to follow in a power emergency. Switches and panels also need to be tested on a regular basis, as directed in your owner's manual. And be sure that all switches (both interior and exterior) are housed in an approved enclosure box.

Pull-cord starter

Smaller portable generators often use pull cords instead of electric starters.

Anatomy of a Portable Backup Generator

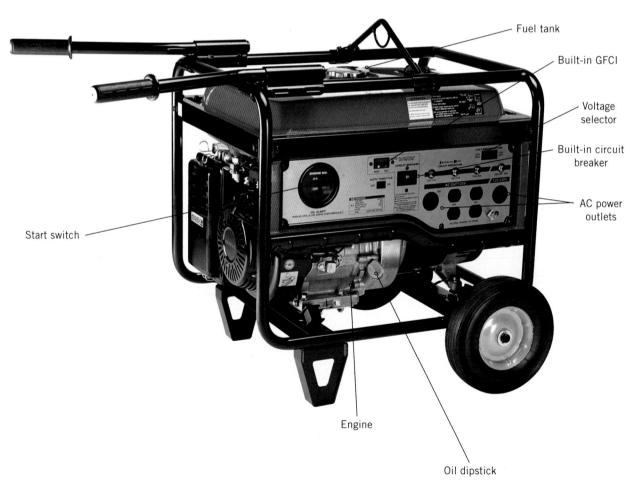

Fuel tank

Built-in GFCI

Voltage selector

Built-in circuit breaker

AC power outlets

Start switch

Engine

Oil dipstick

Portable generators use small gasoline engines to generate power. A built-in electronics panel sets current to AC or DC and the correct voltage. Most models will also include a built-in circuit breaker to protect the generator from damage in the event it is connected to too many loads. Better models include features such as built-in GFCI protection. Larger portable generators may also feature electric starter motors and batteries for push-button starts.

Operating a Manual System During an Outage

Plug the generator in at the inlet box. Make sure the other end of the generator's outlet cord is plugged into the appropriate outlet on the generator (120-volt or 120/240-volt AC) and the generator is switched to the appropriate voltage setting.

Start the generator with the pull cord or electric starter (if your generator has one). Let the generator run for several minutes before flipping the transfer switch.

Flip the manual transfer switch. Begin turning on loads one at a time by flipping breakers on, starting with the ones that power essential equipment. Do not overload the generator or the switch, and do not run the generator at or near full capacity for more than 30 minutes at a time.

Maintaining & Operating an Automatic Standby Generator

If you choose to spend the money and install a dedicated standby generator of 10,000 watts or more and operate it through an automatic transfer switch or panel, you won't need to lift a hand when your utility power goes out. The system kicks in by itself. However, you should follow the manufacturer's instructions for testing the system, changing the oil, and running the motor periodically.

Installing a Transfer Switch

A transfer switch is installed next to the main service panel to override the normal electrical service with power from a backup generator during a power outage. Manual transfer switches require an operator to change the power source, while automatic switches detect the loss of power, start the backup generator, and switch over to the backup power feed. Because the amount of electricity created by a backup generator is not adequate to power all of the electrical circuits in your house, you'll need to designate a few selected circuits to get backup current.

Note: This project requires a permit and inspection of all work.

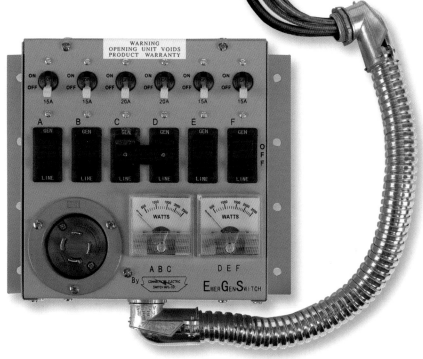

A manual transfer switch connects emergency circuits in your main panel to a standby generator.

TOOLS & MATERIALS

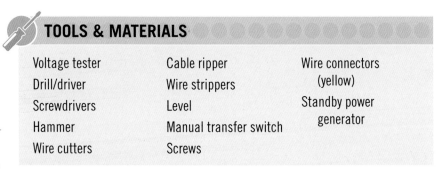

Voltage tester	Cable ripper	Wire connectors (yellow)
Drill/driver	Wire strippers	
Screwdrivers	Level	Standby power generator
Hammer	Manual transfer switch	
Wire cutters	Screws	

One flip of a switch reassigns the power source for each critical circuit so your backup generator can keep your refrigerator, freezer, and important lights running during a utility power outage.

SELECTING BACKUP CIRCUITS

Before you purchase a backup generator, determine which loads you will want to power from your generator in the event of a power loss. Generally you will want to power your refrigerator, freezer, and maybe a few lights. Add up the running wattage ratings of the appliances you will power up to determine how large your backup generator needs to be. Because the startup wattage of many appliances is higher than the running wattage, avoid starting all circuits at the same time—it can cause an overload situation with your generator. Here are some approximate running wattage guidelines:

- Refrigerator: 750 watts
- Forced air furnace: 1,100 to 1,500 watts
- Incandescent lights: 60 watts per bulb (CFL and LED lights use less wattage)
- Sump pump: 800 to 1,000 watts
- Well pump: 2,000 to 5,000 watts
- Garage door opener: 550 to 1,100 watts
- Television: 300 watts

Add the wattage values of all the loads you want to power, and multiply the sum by 1.25. This will give you the minimum wattage your generator must produce. Portable standby generators typically output 5,000 to 7,500 watts. Most larger, stationary generators can output 10,000 to 20,000 watts (10 to 20 kilowatts).

 # How to Install a Manual Transfer Switch

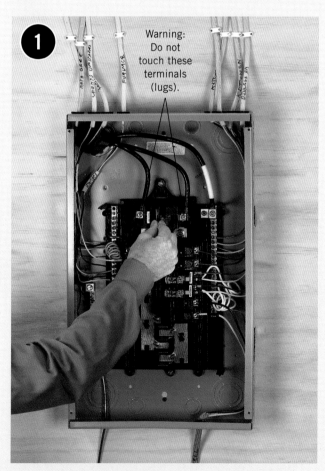

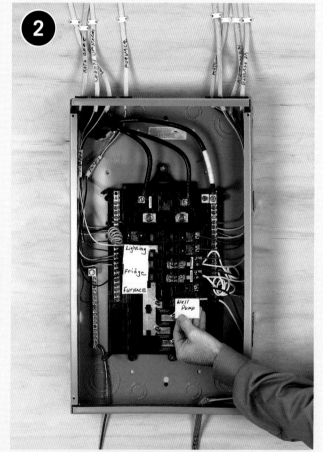

Turn off the main power breaker in your electrical service panel. CAUTION: The service wires and terminals (lugs) that feed the main breaker remain live even when the main breaker is off.

Determine which household circuits are critical for emergency usage during a power outage. Typically this will include the refrigerator, freezer, furnace, and at least one light or small-appliance circuit. *(continued)*

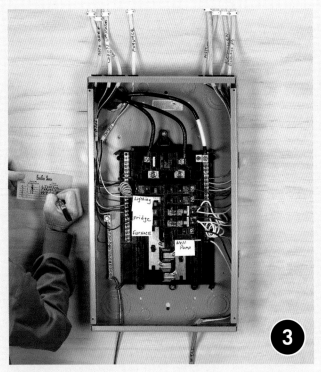

Match your critical circuits with circuit inlet on your prewired transfer switch. Try to balance the load as best you can in the transfer switch: For example, if your refrigerator is on the leftmost switch circuit, connect your freezer to the circuit farthest to the right. Double-pole (240-volt) circuits will require two 120-volt circuit connections. Also make sure that 15-amp and 20-amp circuits are not mismatched with one another.

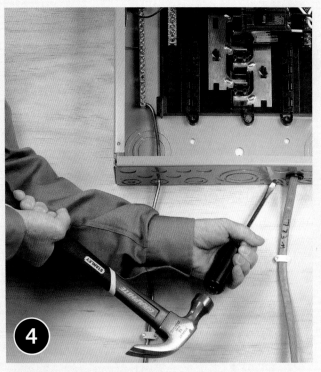

Select and remove a knockout at the bottom of the main service panel box. Make sure to choose a knockout that is sized to match the connector on the flexible conduit coming from the transfer switch.

Feed the wires from the transfer switch into the knockout hole, taking care not to damage the insulation. You will note that each wire is labeled according to which circuit in the switch box it feeds.

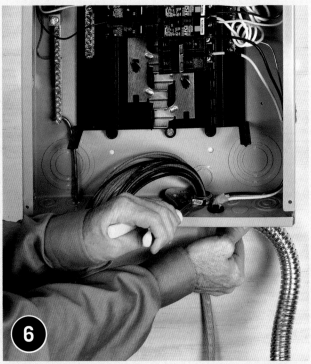

Secure the flexible conduit from the switch box to the main service panel using a locknut and a bushing where required.

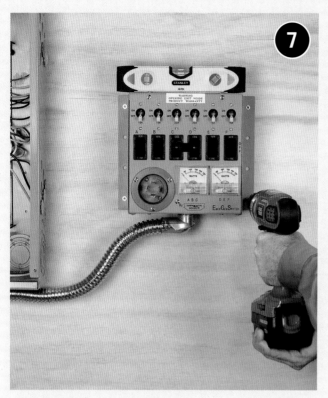

Attach the transfer switch box to the wall so the closer edge is about 18" away from the center of the main service panel. Use whichever connectors make sense for your wall type.

Remove the breaker for the first critical circuit from the main service panel box, and disconnect the hot wire lead from the lug on the breaker.

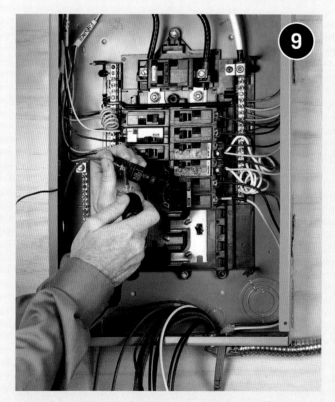

Locate the red wire for the switch box circuit that corresponds to the circuit you've disconnected. Attach the red wire to the breaker you've just removed, and then reinstall the breaker.

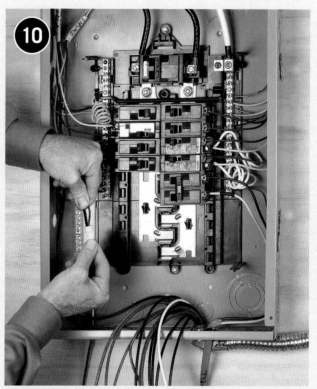

Locate the black wire from the same transfer switch circuit, and twist it together with the old feed wire, using a yellow wire connector. Tuck the wires neatly out of the way at the edges of the box. Proceed to the next circuit, and repeat the process. *(continued)*

If any of your critical circuits are 240-volt circuits, attach the red leads from the two transfer switch circuits to the double-pole breaker. The two circuits originating in the transfer switch should be next to one another, and their switches should be connected with a handle tie. If you have no 240-volt circuits, you may remove the preattached handle tie and use the circuits individually.

Once you have made all circuit connections, attach the white neutral wire from the transfer switch to an opening in the neutral terminal bar of the main service panel.

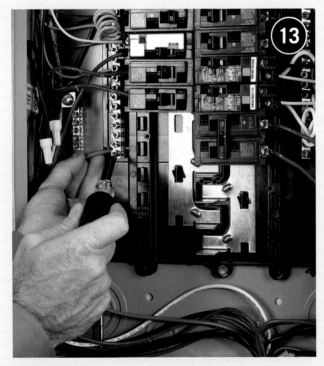

Attach the green ground wire from the transfer switch to an open port on the grounding bar in your main service panel. This should complete the installation of the transfer switch. Replace the cover on the service panel box, and make sure to fill in the circuit map on your switch box.

Begin testing the transfer switch by making sure all of the switches on it are set to the LINE setting. The power should still be OFF at the main panel breakers.

Make sure your standby generator is operating properly and has been installed professionally. See page 75 for information on choosing a generator that is sized appropriately for your needs.

Before turning your generator on, attach the power cord from the generator to the switch box. Never attach or detach a generator cord with the generator running. Turn your standby power generator on, and let it run for a minute or two.

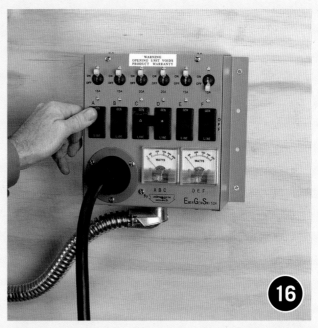

Flip each circuit switch on the transfer switch box to GEN, one at a time. Try to maintain balance by moving back and forth from circuits on the left and right side. Do not turn all circuits on at the same time. Observe the onboard wattage meters as you engage each circuit, and try to keep the wattage levels in balance. When you have completed testing the switch, turn the switches back to LINE, and then shut off your generator.

Outbuildings

Nothing improves the convenience and usefulness of an outbuilding more than electrifying it. Running a new underground circuit from your house to an outbuilding lets you add receptacles and light fixtures both inside the outbuilding and on its exterior. If you run power to an outbuilding, you are required to install at least one receptacle.

Adding one or two 120-volt circuits is not complicated, but every aspect of the project is strictly governed by local building codes. Therefore, once you've mapped out the job and have a good idea of what's involved, visit your local building department to discuss your plans and obtain a permit for the work.

This project demonstrates standard techniques for running a circuit cable from the house exterior to a shed, plus the wiring and installation of devices inside the shed.

First, determine how much current you will need. For basic electrical needs, such as powering a standard light fixture and small appliances or power tools, a 120-volt, 20-amp circuit should be sufficient. A small workshop may require one or two 120-volt, 20-amp circuits. If you need any 240-volt circuits or more than two 120-volt, 20-amp circuits, you will need to install at least a 60-amp subpanel with appropriate feeder wires. Installing a subpanel in an outbuilding is similar to installing one inside your home, but there are some important differences.

You may use #12 copper wire for one 120-volt, 20-amp circuit. Use #10 copper wire for two 120-volt, 20-amp circuits. Also, if the shed is more than 150 feet away from the house, you may need heavier-gauge cable to account for voltage drop.

Most importantly, don't forget to call before you dig. Have all utility and service lines on your property marked even before you make serious project plans. This is critical for your safety of course, and it may affect where you can run the circuit cable.

Adding an electrical circuit to an outbuilding such as this greatly expands the activities the building will support and is also a great benefit for home security.

Spray paint

Trenching shovel (4"-wide blade)

4" metal junction box

Metal L-fittings (2) and conduit nipple for conduit

Wood screws

Conduit with watertight threaded and compression fittings

Wrenches

Hacksaw

90° sweeps for conduit (2)

Plastic conduit bushings (2)

Pipe straps

Silicone caulk and caulk gun

Double-gang boxes, metal (2)

One exterior receptacle box (with cover)

One 20-amp weather-resistant receptacle

One 20-amp receptacle

Single-pole switches (2)

Interior ceiling light fixture and metal fixture box

Exterior motion-detector fixture and plastic fixture box

EMT metal conduit and fittings for inside the shed

Utility knife

UF two-wire cable (12 gauge)

THNN wire (12 gauge)

20-amp GFCI circuit breaker

Wire strippers

Pliers

Screwdrivers

Wire connectors

Hand tamper

Schedule 80 conduit

Eye protection

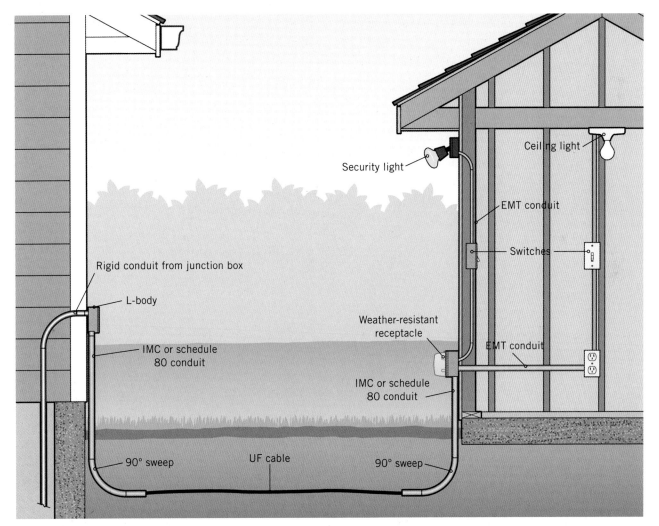

A basic outdoor circuit starts with a waterproof fitting at the house wall connected to a junction box inside. The underground circuit cable—rated UF (underground feeder)—runs in a 12"-deep trench and is protected from exposure at both ends by metal or PVC conduit. Inside the shed, standard NM cable runs through metal conduit to protect it from damage (not necessary if you will be adding interior wallcoverings). All receptacles must have GFCI protection; this is provided by a GFCI circuit breaker.

Identify the circuit's exit point at the house and entry point at the shed and mark them. Mark the path of the trench between the exit and entry points using spray paint. Make the route as direct as possible. Dig the trench to the depth required by local code (typically 12" deep for a GFCI-protected circuit) using a narrow trenching shovel.

From outside, drill a hole through the exterior wall and the rim joist at the exit point for the cable (you'll probably need to install a bit extender or an extra-long bit in your drill). Make the hole just large enough to accommodate the L-body conduit fitting and conduit nipple.

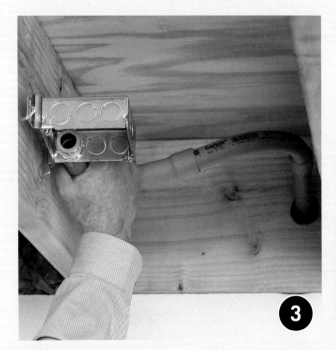

Assemble the conduit and junction box fittings that will penetrate the wall. Here, we attached a 12" piece of ¾" PVC conduit and a sweep to a metal junction box with a compression fitting and then inserted the conduit into the hole drilled in the rim joist. The junction box is attached to the floor joist.

From outside, seal the hole around the conduit with expandable spray foam or caulk, and then attach the free end of the conduit to the back of a waterproof L-body fitting. Mount the L-body fitting to the house exterior with the open end facing downward.

Cut a length of PVC conduit to extend from the L-fitting down into the trench using a hacksaw. Deburr the cut edges of the conduit. Secure the conduit to the L-fitting, and then attach a 90° sweep to the bottom end of the conduit using compression fittings. Anchor the conduit to the wall with a corrosion-resistant pipe strap.

Inside the shed, drill a ¾"-diameter hole in the shed wall. On the interior of the shed, mount a junction box with a knock-out removed to allow the cable to enter through the hole. On the exterior side directly above the end of the UF trench, mount an exterior-rated receptacle box with cover. The plan (and your plan may differ) is to bring power into the shed through the hole in the wall behind the exterior receptacle.

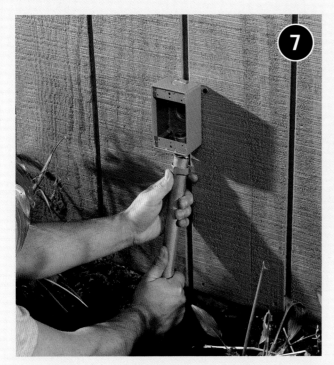

Run conduit from the exterior box down into the trench. Fasten the conduit to the outbuilding with a strap. Add a 90° sweep and bushing, as before. Secure the conduit to the box with an offset fitting. Anchor the conduit with pipe straps, and seal the entry hole with caulk.

Run UF cable from the house to the outbuilding. Feed one end of the UF circuit cable up through the sweep and conduit and into the L-fitting at the house (the back or side of the fitting is removable to facilitate cabling). Run the cable through the wall and into the junction box, leaving at least 12" of extra cable at the end. *(continued)*

Lay the UF cable into the trench, making sure it is not twisted and will not contact any sharp objects. Roll out the cable, and then feed the other end of the cable up through the conduit and into the receptacle box in the shed, leaving 12" of slack.

Inside the outbuilding, install the remaining boxes for the other switches, receptacles, and lights. With the exception of plastic receptacle boxes for exterior exposure, use metal boxes if you will be connecting the boxes with metal conduit.

Connect the electrical boxes with conduit and fittings. Inside the outbuilding, you may use inexpensive steel EMT to connect the receptacle, switch, and fixture boxes. Once you've planned your circuit routes, start by attaching couplings to all of the boxes.

Cut a length of conduit to fit between the coupling and the next box or fitting in the run. If necessary, drill holes for the conduit through the centers of the wall studs. Attach the conduit to the fitting that you attached to the first box.

13

If you are surface-mounting the conduit or running it up or down next to wall studs, secure it with straps no more than 3 ft. apart. Use elbow fittings for 90° turns and setscrew couplings for joining straight lengths as needed. Make holes through the wall studs only as large as necessary to feed the conduit through.

14

THNN wire

Measure to find how much wire you'll need for each run, and cut pieces of THHN wire that are 1 or 2 feet longer than the measurements. Before making L-turns with the conduit, feed the wire through the first conduit run.

15

Feed the other ends of the wires into the next box or fitting in line. It is much easier to feed wire into 45° and 90° elbows if they have not been attached to the conduit yet. Continue feeding wire into the conduit and fitting until you have reached the next box in line.

16

LIGHT

Once you've reached the next box in line, coil the ends of the wires and repeat the process with new wire for the next run. Keep working until all of the wire is run and all of the conduit and fittings are installed and secured. If you are running multiple feed wires into a single box, write the origin or destination on a piece of masking tape and stick it to each wire end. *(continued)*

17

NOTE: Your code may require an in-use rated receptacle box cover (see page 51).

Make the wiring connections at the receptacles. Connect the receptacles with pigtails, including grounding pigtails for the receptacles and the metal boxes. Install the receptacles and cover plates.

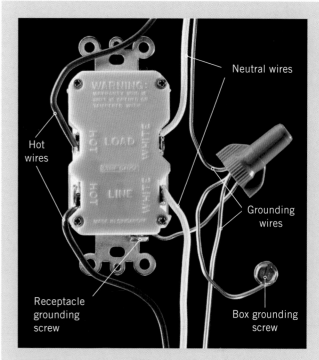

Neutral wires

Hot wires

Grounding wires

Receptacle grounding screw

Box grounding screw

VARIATION: Installing a GFCI breaker for the new circuit at the main service panel is the best way to protect the circuit and allows you to use regular receptacles in the building, but an alternative that is allowed in many areas is to run the service into a GFCI-protected receptacle and then wire the other devices on the circuit in series. If you use this approach, only the initial receptacle needs to be a GFCI receptacle; however, the underground circuit cable will need to be at least 24" deep.

18

Run service from the last receptacle to the switch box for the light fixture or fixtures. (If you anticipate a lot of load on the circuit, you should probably run a separate circuit for the lights). Twist the white neutral leads and grounding leads together and cap them. Attach the black wires to the appropriate switches. Install the switches and cover plate.

19

Install the light fixtures. For this shed, we installed a caged ceiling light inside the shed and a motion-detector security light on the exterior side.

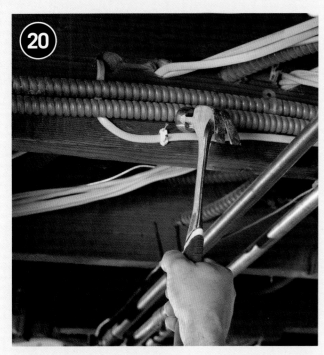

Run NM cable from the electrical box in the house at the start of the new circuit to the main service panel. Use cable staples if you are running the cable in floor joist cavities. If the cable is mounted to the bottom of the floor joists or will be exposed, run it through conduit.

At the service panel, feed the NM cable in through a cable clamp. Arrange for your final electrical inspection before you install the breaker. Then attach the wires to a new GFCI circuit breaker, and install the breaker in an empty slot. Label the new circuit on the circuit map.

Turn on the new circuit, and test all of the receptacles and fixtures. If any of the fixtures or receptacles is not getting power, check the connections first, and then test the receptacle or switch for continuity with a multimeter. Backfill the trench.

Bathroom Exhaust Fans

Most exhaust fans are installed in the center of the bathroom ceiling or over the toilet area. A fan installed over the tub or shower area must be rated for use in wet areas. You can usually wire a fan that just has a light fixture into a main bathroom lighting circuit (but not into a dedicated bathroom receptacle circuit). Units with built-in heat lamps or blowers require separate circuits. Extending a branch circuit or adding a new branch to install new receptacles, lights, switches, or equipment requires a permit. Check with the electrical inspector before starting such projects.

If the fan you choose doesn't come with a mounting kit, purchase one separately. A mounting kit should include an exhaust hose (duct), a vent tailpiece, and an exterior terminal.

Three common places to terminate the exhaust are the roof, a soffit, or a sidewall. The instructions in this book are for a shingle roof covering. You should have a roofer install the exhaust termination if you have any other roofing material or if you are not comfortable walking on your roof.

A soffit exhaust involves routing the duct to a soffit (roof overhang) where it is connected to a terminal that directs the exhaust outside. While soffit exhausts are allowed, they are not recommended, because the moisture can be drawn back into the attic through the soffit vents. Check with the exhaust fan manufacturer for instructions about how to run and terminate the exhaust duct and to determine the required duct diameter and maximum length.

To prevent moisture damage, always terminate the exhaust duct outside your home—never into your attic.

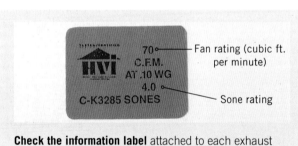

Check the information label attached to each exhaust fan unit. The minimum rating is 50 CFM; larger baths may need up to 100 CFM. The sone rating refers to quietness rated on a scale of 1 to 7; quieter is lower.

TOOLS & MATERIALS

Drill	NM cable (14/2, 14/3)
Jigsaw	Cable clamp
Combination tool	Hose clamps
Screwdrivers	Pipe insulation
Caulk gun	Roofing cement
Reciprocating saw	Self-sealing roofing nails
Pry bar	Shingles
Screws	Wire connectors
Double-gang retrofit electrical box	Switch and timer
	Eye protection

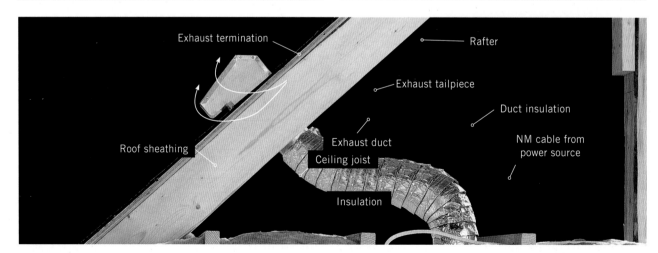

Bathroom exhaust fans must be exhausted to the outdoors, either through the roof or through a wall. Three-inch-diameter flexible duct is not allowed for bathroom exhaust fans. Four-inch-diameter flexible duct is allowed for up to 56 feet without bends for 50 cfm fans. Use fan manufacturer's installation instructions for duct type and length. Use insulated duct in cold climates to reduce moisture damage from condensation on the duct.

How to Install a Bathroom Exhaust Fan

Position the fan unit against a ceiling joist. Outline the fan onto the ceiling surface. Remove the unit, drill pilot holes at the corners of the outline, and cut out the area with a jigsaw or drywall saw.

Remove the grille from the fan unit, and then position the unit against the joist with the edge recessed ¼" from the finished surface of the ceiling (so the grille can be flush mounted). Attach the unit to the joist using drywall screws.

VARIATION: For fans with heaters or light fixtures, some manufacturers recommend using 2× lumber to build dams between the ceiling joists to keep the insulation at least 6" away from the fan unit.

Switch box location

Mark and cut an opening for a double-gang box on the wall next to the latch side of the bathroom door, and then run a 14/3 NM cable from the switch cutout to the fan unit. Run a 14/2 NM cable from the power source to the cutout.

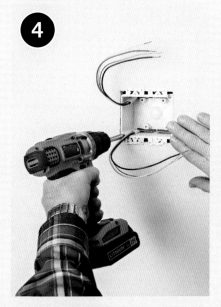

Strip 10" of sheathing from the ends of the cables, and then feed the cables into a double-gang retrofit switch box so at least ½" of sheathing extends into the box. Clamp the cables in place. Tighten the mounting screws until the box is secure.

Strip 10" of sheathing from the end of the cable at the unit, and then attach a cable clamp to the cable. Insert the cable into the fan unit. From the inside of the unit, screw a locknut onto the threaded end of the clamp.

(continued)

6

Mark the exit location in the roof next to a rafter for the exhaust duct. Drill a pilot hole, and then saw through the sheathing and roofing material with a reciprocating saw to make the cutout for the exhaust tailpiece.

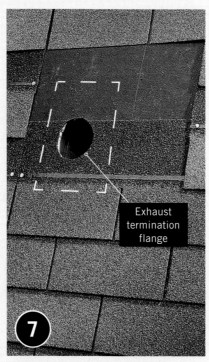

7

Exhaust termination flange

Remove a section of shingles from around the cutout, leaving the roofing paper intact. Remove enough shingles to create an exposed area that is at least the size of the exhaust termination flange.

8

Exhaust tailpiece

Attach a hose clamp to the rafter next to the roof cutout about 1" below the roof sheathing (top). Insert the exhaust tailpiece into the cutout and through the hose clamp, and then tighten the clamp screw (bottom).

9

Slide one end of the exhaust duct over the tailpiece, and slide the other end over the outlet on the fan unit. Slip hose clamps or straps around each end of the duct, and tighten the clamps. Wrap the exhaust duct with pipe insulation. Insulation prevents moist air inside the duct from condensing and dripping down into the fan motor.

10

Apply roofing cement to the bottom of the exhaust termination flange, and then slide the termination over the tailpiece. Nail the termination flange in place with self-sealing roofing nails, and then patch in shingles around the cover.

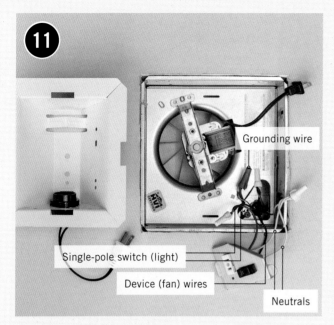

11

Grounding wire

Single-pole switch (light)

Device (fan) wires

Neutrals

Turn power off and test for power. Make the following wire connections at the fan unit: the black circuit wire from the timer to the wire lead for the fan motor; the red circuit wire from the single-pole switch (see step 14) to the wire lead for the light fixture in the unit; the white neutral circuit wire to the neutral wire lead; the circuit grounding wire to the grounding lead on the fan unit. Make all connections with wire connectors. Attach the cover plate over the unit when the wiring is completed.

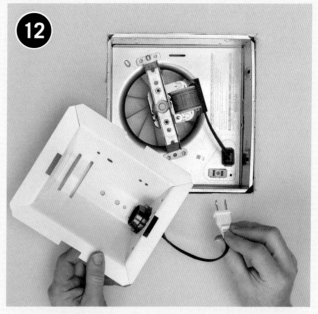

12

Connect the light plug to the built-in receptacle on the wire connection box. Attach the fan grille to the frame using the mounting clips included with the fan kit.

NOTE: If you removed the wall and ceiling surfaces for the installation, install new surfaces before completing this step.

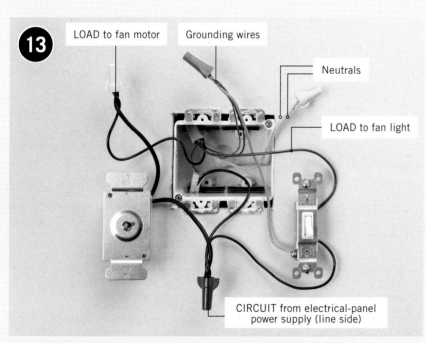

13

LOAD to fan motor   Grounding wires

Neutrals

LOAD to fan light

CIRCUIT from electrical-panel power supply (line side)

Turn power off and test for power. At the switch box, add black pigtail wires to one screw terminal on the timer and to one screw terminal on the single-pole switch; add a green grounding pigtail to the groundling screw on the switch. Make the following wire connections: the black circuit wire from the power source to the black pigtail wires; the black circuit wire from the exhaust fan to the remaining screw on the timer; the red circuit wire from the exhaust fan to the remaining screw on the switch. Join the white wires with a wire connector. Join the grounding wires with a green wire connector.

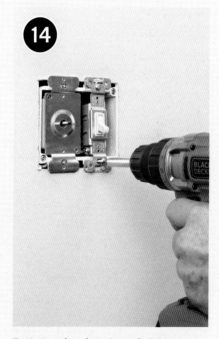

14

Tuck the wires into the switch box, and then attach the switch and timer to the box. Attach the cover plate and timer dial. Turn on the power.

"Smart" Thermostats

Advancing digital technology has impacted just about every aspect of our home systems, including the thermostat. What used to be a fairly simple device (often containing small amounts of mercury) has been replaced in most homes by "smart" thermostats.

The good news, for those who may be slower to embrace new technology, is that how you install a "smart" thermostat is really not that different from the old way of doing it. Each unit has four to eight 20-gauge copper wires that are color coded. You simply mount the thermostat body in the old location and make like-to-like connections to the thermostat cable coming through the wall and leading to your furnace and air conditioner. It is a relatively simple remove-and-replace procedure as long as your existing cable matches the colored wire combination in your new unit. If the cable does not match the device you will need to upgrade it, starting from the appliances and fishing the cable through to the thermostat location.

Once you have made the connections and mounted the thermostat to the wall, you simply need to download the correct app to your smartphone and follow the setup and usage directions. Quality smart thermostats may be expensive. But once they are set up they typically do not require any monthly usage fees.

Programmable thermostats contain sophisticated circuitry that allows you to set the heating and cooling systems in your house to adjust automatically at set times of the day. Replacing a manual thermostat with a programmable model is a relatively simple job that can have big payback on heating and cooling energy savings.

Internet-connected thermostats can be controlled by using an app installed on your smart phone. They are relatively expensive, starting around $200. Installation is similar to older programmable thermostats—the real trick is making sure you have the right number of colored wires and that you connect them properly.

How to Upgrade to a Programmable Thermostat

Start by removing the existing thermostat. Turn off the power to the furnace at the main service panel, and test for power. Then remove the thermostat cover.

The body of the thermostat is held to a wall plate with screws. Remove these screws, and pull the body away from the wall plate. Set the body aside.

The low-voltage wires that power the thermostat are held by screw terminals to the mounting plate. Do not remove the wires until you label them with tape according to the letter printed on the terminal to which each wire is attached.

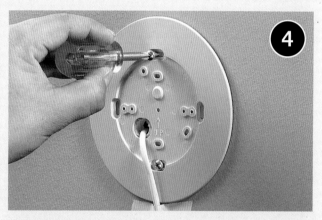

Once all the wires are labeled and removed from the mounting plate, tape the cable that holds these wires to the wall to keep it from falling back into the wall. Then unscrew the mounting plate and set it aside.

Position the new thermostat base on the wall, and guide the wires through the central opening. Screw the base to the wall using wall anchors if necessary.

Check the manufacturer's instructions to establish the correct terminal for each low-voltage wire. Then connect the wires to these terminals, making sure each screw is secure.

Programmable thermostats require batteries to store the programs so they won't disappear if the power goes out in a storm. Make sure to install batteries before you snap the thermostat cover in place. Program the new unit to fit your needs, and then turn on the power to the furnace.

How to Install an Internet-based Thermostat

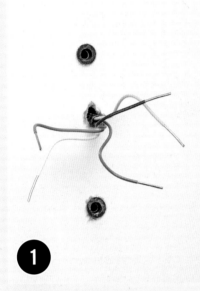

1

Remove the old thermostat and compare the color and quantity of wires from the wall to the instructions that came with the new thermostat. If they are not compatible you will have to purchase new thermostat wire that is compatible with the new unit and run it from the furnace/air conditioner to the thermostat location.

2

Thread the thermostat wires through the access hole in the thermostat base plate and attach the plate and mounting ring (optional) to the wall.

3

Attach the color coded wires to the correct terminal on the base plate unit according to the manufacturer's directions.

4

Make sure a new battery of the correct size is installed in the thermostat and then snap the cover/sensor onto the base plate. To engage the new thermostat for internet usage you will need to download an app for your smart phone from the thermostat manufacturer's website.

Troubleshooting & Repairs

Running new circuits and hooking up new fixtures are fairly predictable projects when it comes to estimating time and expense. This is less true with repairing problems in your system and fixtures. In some cases, a repair is as simple as opening an electrical box, spotting a loose wire connector and remaking the connection. But there are also times when fixing a dead circuit or device is a highly frustrating proposition. Such cases are almost always caused by tricky diagnostic challenges. Wires are hidden behind walls and there very often are no visual clues to system breakdowns. So essentially, minimizing repair frustration boils down to learning to deploy logical, systematic diagnostics. Educated troubleshooting, you could say.

In this project you'll learn how to use the most important diagnostic tool in any electrician's toolkit: the multimeter. These handy devices come in a dizzying array of types and qualities, but for diagnostic purposes they are used to take readings for current (amperage), voltage and continuity (whether an electrical path is open or closed). Once you learn the basics of operating a multimeter, you can enlist it in a logical, deductive manner to track down the source of a wiring problem. Once located, correcting the problem is usually very simple.

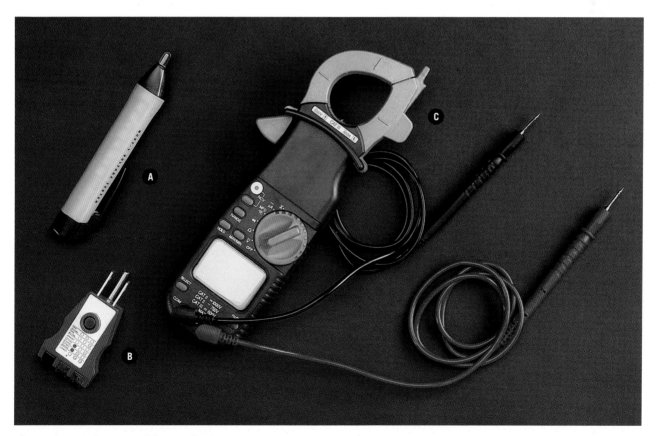

Diagnostic tools for home wiring use include: Touchless circuit tester (A) to safely check wires for current and confirm that circuits are dead; Plug-in tester (B) to check receptacles for correct polarity, grounding and circuit protection; Multimeter (C) to measure AC/DC voltage, AC/DC current, resistance, capacitance, frequency and duty cycle (model shown is an auto-ranging digital multimeter with clamp-on jaws that measure through sheathing and wire insulation).

Multimeters

Multimeters are nearly indispensible diagnostic tools for doing intermediate to advanced level electrical work (as well as automotive and electronics repair). They are used to measure voltage, current (amperage) and a few other conditions such as continuity, capacitance and frequency. For your home electrical system, by far the most used feature of a multimeter is testing voltage and current, although there are occasions where testing for resistance is needed. Among professional electricians, the most common and widely used multimeters have a clamp-on ammeter that measures current through the wire insulation so you don't have to disconnect the circuit and expose bare wire. Most clamp-on multimeters also are fitted with insertible probes with which you can measure voltage and continuity in the traditional way. An example of a clamp-on multimeter can be seen on the next page. Among homeowners, however, the most common multimeters these days are digital, auto-ranging tools that use probes or alligator clamps at the ends of wire leads for diagnostic work. Older multimeters that do not have autoranging capability must be pre-set to estimated calibration levels before use. Non-digital multimeters or ammeters usually have a dial gauge that gives readouts. These tools are somewhat more difficult to use and are less precise. Considering that digital, autoranging multimeters can be found for just a few dollars (the top of the line models cost over $100) there is really no good reason not to replace your old device with one that resembles the tools seen on these pages.

TIME TO REPLACE THAT NEON TESTER

Neon circuit testers are inexpensive and easy to use (if the light glows the circuit is hot), but they are less sensitive than multimeters and can be unsafe. In some cases, neon testers won't detect the presence of lower voltage in a circuit. This can lead you to believe that a circuit is shut off when it is not—a dangerous mistake. The small probes on a neon circuit tester also force you to get too close to live terminals and wires. For the most reliable readings, buy and learn to use a multimeter. At the very least, switch to a touchless tester.

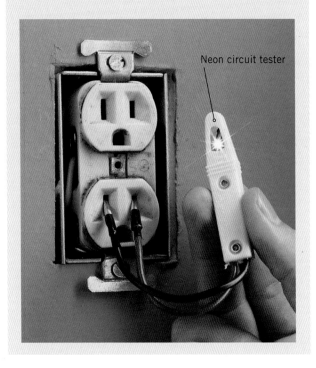

Neon circuit tester

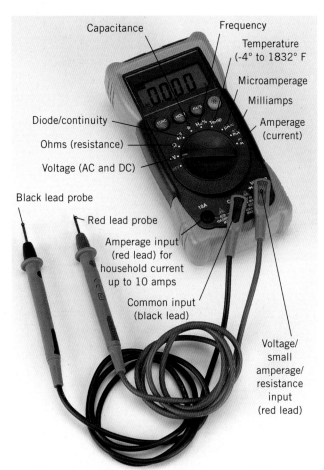

Capacitance

Frequency

Temperature (-4° to 1832° F

Microamperage

Milliamps

Diode/continuity

Amperage (current)

Ohms (resistance)

Voltage (AC and DC)

Black lead probe

Red lead probe

Amperage input (red lead) for household current up to 10 amps

Common input (black lead)

Voltage/ small amperage/ resistance input (red lead)

A digital, autoranging multimeter must be adjusted to the proper setting for the reading you want to take. The probe leads also must be inserted into the correct inlet at the bottom of the tool. Inserting the red lead into the incorrect inlet can cause the tool to trip an internal fuse. Study your owner's manual carefully before using any tool.

How to Measure Current

Create access to the wires you need to test. In most cases this requires that you remove the cover to an electrical service panel or an electrical box (inset).

Set the multimeter to test for amperage (current is measured in amperes or amps). On some multimeters you need to select between amperage settings that are above or below 40 amps. Use the rated amperage of the circuit as a guide (amperage is printed on the circuit breaker switch).

Clamp the jaws of a clamp-on multimeter onto the conductor or one of the conductors (if more than one) leading to a circuit breaker. If you are using a non-clamping multimeter, touch one probe to the screw terminal where the hot lead is attached to the breaker and touch the other probe to the neutral terminal bar. The readout on your meter is the amount of current flowing in that circuit.

 TAKING MEASUREMENTS AT A RECEPTACLE

You may use a multimeter to measure for voltage at a wall receptacle. Regardless of whether the outlet is in service, if it is live you will get a voltage reading in the approximate range of the receptacle rating—here, 120 volts. To detect live current, measured in amps, the receptacle must be in use, with an appliance drawing from it. Taking an amperage reading in such an instance will only yield the amount of current being drawn, which is a factor of the appliance, not the circuit capacity.

How to Measure Voltage

To measure voltage using the multimeter, you will have to use the two probes provided with the multimeter and have access to a live terminal or slot as well as a grounded terminal or slot. If your meter has probe holders at the top, snap the probes into them. They are like extra hands.

Turn the multimeter to the VAC setting to measure AC voltage that is found in your house. Set the multimeter to VDC if measuring DC voltage, such as in a car or a battery-fed device. On some multimeters, like the one above, you select "V" for voltage then change between AC and DC with the"FUNC" button.

To measure the AC voltage, place one probe on a grounded surface, such as the metallic junction box or the bare ground wire. Place the other probe on the hot screw terminal or into the receptacle slot associated with the hot wire. The voltage readout should be in the range of 120 volts, plus or minus 5 volts (usually 120 volts in a residence in the US).

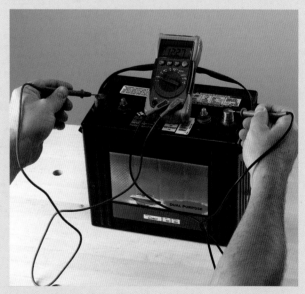

240 VOLTS. You can also measure voltage across the two hot leads to determine if you have 240 volts. This can be done at your range receptacle, dryer receptacle, or any other 240-volt receptacle. Place one probe in one of the small slots and the other probe in the other slot directly across from it. The voltage should read 240 volts, plus or minus 5 volts.

DC VOLTAGE. When testing DC voltage, such as in a car battery, you can measure exactly the same way as for AC as long as the meter is set to the DC function. For more accurate results, test the voltage while the battery is in use.

How to Test for Continuity

Continuity is a condition in a circuit where the conductors form an unbroken pathway through which current may flow. When measuring for continuity, always make sure there is no power present on the circuit you are testing or damage may occur to the meter. You can also measure the resistance in this mode as well.

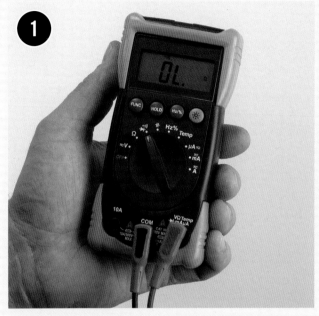

The setting for continuity is an "audible" or diode symbol display on the dial. Select this setting.

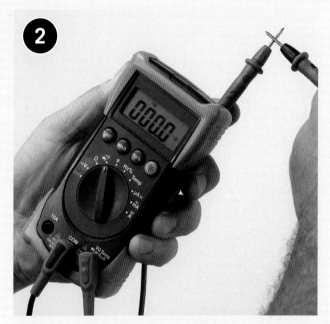

Verify that the continuity tester is functional by touching the two probes together. You should hear an alert sound and/or see a reading of zero ohms (Ohms is a value of the resistance to current flow).

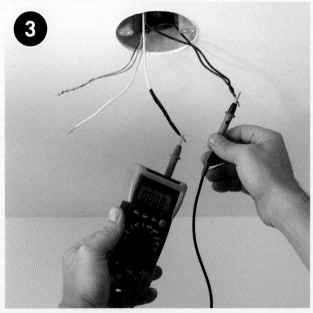

To test a circuit, touch one probe to one of the wires on a given circuit and the other to the second wire of the circuit. If you hear an audible sound or read a value of resistance other than zero, you have a complete or unbroken path for current to flow.

 ## HOW TO TEST A 3-WAY SWITCH

Remove the switch from the circuit and place one of the probes onto the common terminal and the other probe onto one of the other two terminals used for the traveler wires. If the meter indicates infinity ohms or there is no sound, flip the switch and if it is in working order the meter should read zero ohms or emit an audible sound. It should only work in one direction or the other, not both.

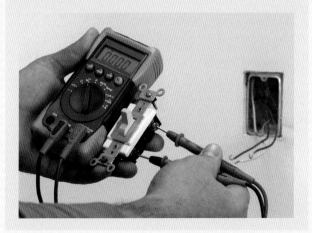

Troubleshooting an Open Neutral

Current and voltage move through the energized (hot) wire, do work in a load (a light, television, blender, etc.), and the current (at zero volts) returns to the utility transformer through the grounded (neutral) wire. If the neutral wire becomes disconnected, electricity might not flow in the circuit, or it might flow in unintended and possibly dangerous ways. There are three different situations in which an open neutral can occur. Each situation presents different indications. These indications will vary based on circuit conditions such as where the open neutral is located, how the circuit is wired, and the loads active on the circuit.

POSSIBLE SYMPTOMS OF AN OPEN NEUTRAL

120-volt, 15- or 20-Amp Light and Receptacle Circuits

- *Circuit Description:* These circuits serve lights and receptacles inside and outside the house. They have one fuse or circuit breaker connected to one hot wire, and one neutral wire.

- *Common Defect Indications:* Some lights and receptacles work, some do not work. Voltage could be detected on the neutral wire. These indications are usually limited to one circuit.

- *Common Defect Locations:* Wire cap connections in light boxes, receptacle boxes, and junction boxes, and in panelboard cabinets; neutral connections at receptacles; splices made outside of boxes anywhere in the circuit.

- *Troubleshooting Techniques:* See next page.

120-volt, 15- or 20-Amp Multi-wire Branch Circuit

- *Circuit Description:* A multi-wire branch circuit consists of two hot wires that share a neutral wire. The most common locations in a house are the circuit serving the dishwasher/disposer and the two 20-amp circuits serving the kitchen countertop receptacles and receptacles in the breakfast room and dining room. There could be other multi-wire branch circuits. Multi-wire branch circuits may usually be identified at the panelboard by a red wire and a black wire that are connected to adjacent circuit breakers sharing one neutral wire in the same cable. Current NEC rules require that all multi-wire branch circuit's circuit breakers be connected with a handle tie, but this is a recent change so circuit breakers in older systems may not be connected with a handle tie.

- *Common Defect Indications:* Lights may glow very dim or very bright and change based on whether other loads are on or off. Appliances may run unusually fast or slow and change based on whether other loads are on or off. Electronics and appliances may not work, or may be damaged or destroyed.

- *Common Defect Locations:* Same as for light and receptacle circuits.

- *Troubleshooting Techniques:* This is best left to experienced electricians. The risk of property damage and personal injury is high.

House Neutral

- *Circuit Description:* The house neutral is the wire from the main service panel to the utility transformer. It is usually uninsulated stranded aluminum.

- *Common Defect Indications:* Same as for multi-wire branch circuits except multiple circuits in the house will probably be affected.

- *Common Defect Locations:* Most likely could be anywhere between the main service panel and the utility transformer.

- *Troubleshooting Techniques:* Leave diagnosis and repair to experienced electricians and the utility company. The risk of property damage and personal injury is very high.

How to Troubleshoot an Open Neutral

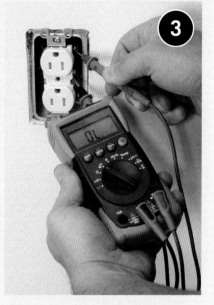

Verify which lights and receptacles are on the circuit by turning the breaker off and by checking for power with a non-contact voltage tester. It is helpful to either draw a map of the house or place some tape on every affected opening.

Start at the outlet nearest to the panel. With the breaker off and using a multimeter, check for continuity between the neutral (white wire) and the ground (bare or green wire). These two wires land at the same point electrically in your electrical panel. If there is an indication of continuity between these two wires, the neutral and ground connections are sound and you should proceed to the next outlet as you move away from the panel.

When you encounter a point at which you read infinity ohms or there is no continuity between the neutral and ground wires, the problem lies within the connections in that box or the box just upstream (toward) the panel from the one you are checking. Sometimes you will see evidence of arcing on the wire cap containing the connection which may include discoloration, or a blackish char near the copper.

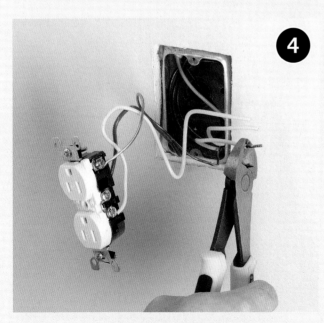

When you have found the problem connection, remove the wire cap and, if it is possible, cut the damaged portions of the wires off and restrip the wires to expose new copper. Line the wire ends up and twist on a new wire cap.

Turn the circuit breaker back on and verify the proper voltage is present at your receptacles by measuring with a multimeter.

Troubleshooting a Short Circuit

Short circuits are a direct connection between the hot or power wire (black or red) and to either the neutral (white) or ground (bare) wire. This connection between the two will cause your circuit breaker or fuse to blow, which should interrupt power to the affected circuit.

Short circuits are a common problem and can usually be solved by taking the following steps. The idea behind electrical troubleshooting is to simplify the circuit by checking it at certain points, in order to narrow down the problem point by process of elimination. Generally, the problem is that there is a bare ground wire touching a hot terminal within a switch or an outlet box. There will usually be a black scorched mark or some sign of an electrical arc where the problem lies.

How to Troubleshoot a Short Circuit

Turn the power off at the affected breaker and verify with a non-contact voltage tester that there is no power present. Unplug everything from the receptacles and turn the lights off on the circuit that is affected.

Using a multimeter set to the ohms or continuity setting, check the wires at the panel. Touch one of the probes to the hot or black wire and the other probe to the ground or bare wire. If the meter rings or indicates a low resistance value, you have a direct short to ground. If the meter does not ring or indicates a high resistance value the circuit is clear. If the meter does not ring, start by turning the switches on one-by-one and re-testing to verify the resistance value. If the meter indicates a low resistance value or a short circuit, the problem is downstream from the switch or within the light fixture itself.

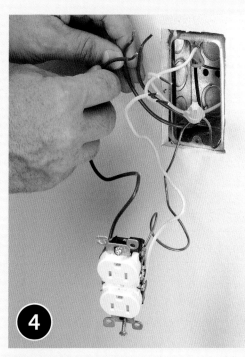

If the meter consistently rings or indicates a low resistive value, you will need to find the electrical box that contains the affected circuit. Choose a box that is convenient to open and preferably in the middle of the run, such as a receptacle. Verify there is no power present by touching all of the wires within the box with a non-contact voltage tester.

If the box you have chosen is in fact in the middle of the run, it will contain at least two cables. Remove the receptacle from the two cables and separate all of the wires.

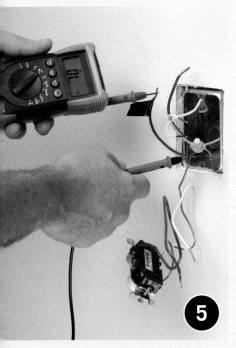

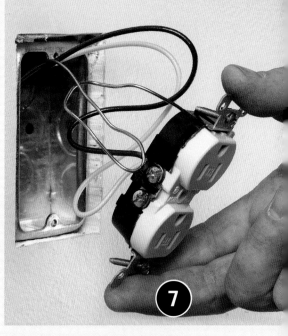

Check the resistance between the black and the ground on both sets of cables. One of the cables should cause the continuity alert to ring and the other should not. Mark the affected one with a piece of black tape and place wire caps over the exposed ends of the black wires.

Check the wires at the panel to see if the short has cleared. If the short is clear, the problem lies down stream from the opened box and it is now safe to turn the breaker back on to help eliminate further problem points. If the short is still present, the problem lies between the opened electrical box and the panel.

Choose another box in the middle of the affected circuit, there by narrowing down the possible problem areas until the short circuit can be positively identified and corrected. When you have discovered the short circuit, verify the wires are still in good shape and repair the connection.

Types of Wall Switches

Wall switches are available in three general types. To reconnect or replace a switch, it is important to identify its type.

Single-pole switches are used to control a set of lights from one location. Three-way switches are used to control a set of lights from two different locations and are always installed in pairs. Four-way switches are used in combination with a pair of three-way switches to control a set of lights from three or more locations.

Identify switch types by counting the screw terminals. Single-pole switches have two screw terminals, three-way switches have three screw terminals, and four-way switches have four. Most switches include a grounding screw terminal, which is identified by its green color.

When replacing a switch, choose a new switch that has the same number of screw terminals as the old one. The location of the screws on the switch body varies depending on the manufacturer, but these differences will not affect the switch operation.

Whenever possible, connect switches using the screw terminals rather than push-in fittings. Some specialty switches (pages 118–119) have wire leads instead of screw terminals. They are connected to circuit wires with wire connectors.

A wall switch is connected to circuit wires with screw terminals or with push-in fittings on the back of the switch. A switch may have a stamped strip gauge that indicates how much insulation must be stripped from the circuit wires to make the connections.

The switch body is attached to a metal mounting strap that allows it to be mounted in an electrical box. Several rating stamps are found on the strap and on the back of the switch. The abbreviation UL or UND. LAB. INC. LIST means that the switch meets the safety standards of the Underwriters Laboratories. Switches also are stamped with maximum voltage and amperage ratings. Standard wall switches are rated 15A or 125V. Voltage ratings of 110, 120, and 125 are considered to be identical for purposes of identification.

For standard wall switch installations, choose a switch that has a wire gauge rating of #12 or #14. For wire systems with solid-core copper wiring, use only switches marked COPPER, CU, or CO/ALR. For aluminum wiring, use only switches marked CO/ALR. Note that while CO/ALR switches and receptacles are approved by the National Electrical Code for use with aluminum wiring, the Consumer Products Safety Commission does not recommend using these. Switches and receptacles marked AL/CU can no longer be used with aluminum wiring, according to the National Electrical Code.

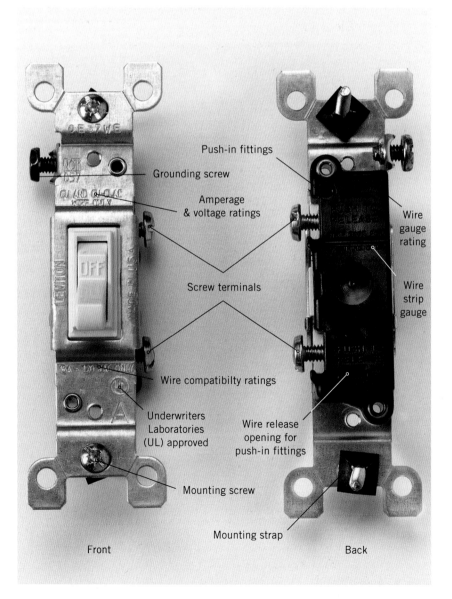

Push-in fittings

Grounding screw

Amperage & voltage ratings

Wire gauge rating

Screw terminals

Wire strip gauge

Wire compatibilty ratings

Underwriters Laboratories (UL) approved

Wire release opening for push-in fittings

Mounting screw

Mounting strap

Front   Back

Single-Pole Wall Switches

A single-pole switch is the most common type of wall switch. It has ON-OFF markings on the switch lever and is used to control a set of lights, an appliance, or a receptacle from a single location. A single-pole switch has two screw terminals and a grounding screw. When installing a single-pole switch, check to make sure the ON marking shows when the switch lever is in the up position.

In a correctly wired single-pole switch, a hot circuit wire is attached to each screw terminal. However, the color and number of wires inside the switch box will vary, depending on the location of the switch along the electrical circuit.

If two cables enter the box, then the switch lies in the middle of the circuit. In this installation, both of the hot wires attached to the switch are black.

If only one cable enters the box, then the switch lies at the end of the circuit. In this installation (sometimes called a switch loop), one of the hot wires is black, but the other hot wire usually is white. A white hot wire should be coded with black tape or paint.

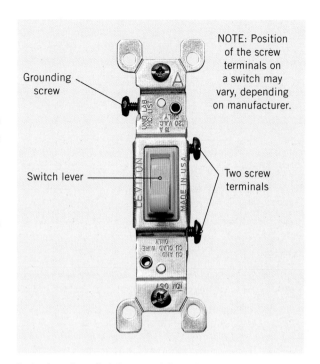

NOTE: Position of the screw terminals on a switch may vary, depending on manufacturer.

Grounding screw

Switch lever

Two screw terminals

A single-pole switch is essentially an interruption in the black power supply wire that is opened or closed with the toggle. Single-pole switches are the simplest of all home wiring switches.

Typical Single-Pole Switch Installations

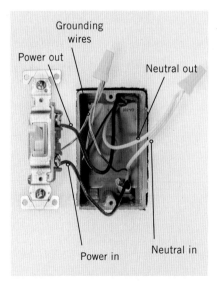

Grounding wires

Power out

Neutral out

Power in

Neutral in

Two cables enter the box when a switch is located in the middle of a circuit. Each cable has a white and a black insulated wire, plus a bare copper grounding wire. The black wires are hot and are connected to the screw terminals on the switch. The white wires are neutral and are joined together with a wire connector. Grounding wires are pigtailed to the switch.

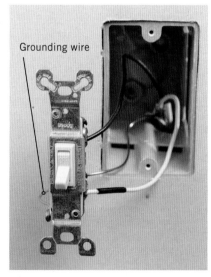

Grounding wire

Old method: One cable enters the box when a switch is located at the end of a circuit. In this installation, both of the insulated wires are hot. The white wire should be labeled with black tape or paint to identify it as a hot wire. The grounding wire is connected to the switch grounding screw.

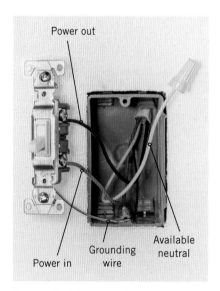

Power out

Available neutral

Power in

Grounding wire

New method: In new switch wiring, the white wire should not supply current to the switched device and a separate neutral wire should be available in the switch box.

Three-Way Wall Switches

Three-way switches have three screw terminals and do not have ON-OFF markings. Three-way switches are always installed in pairs and are used to control a set of lights from two locations.

One of the screw terminals on a three-way switch is darker than the others. This screw is the common screw terminal. The position of the common screw terminal on the switch body may vary, depending on the manufacturer. Before disconnecting a three-way switch, always label the wire that is connected to the common screw terminal. It must be reconnected to the common screw terminal on the new switch.

The two lighter-colored screw terminals on a three-way switch are called the traveler screw terminals. The traveler terminals are interchangeable, so there is no need to label the wires attached to them.

Because three-way switches are installed in pairs, it sometimes is difficult to determine which of the switches is causing a problem. The switch that receives greater use is more likely to fail, but you may need to inspect both switches to find the source of the problem.

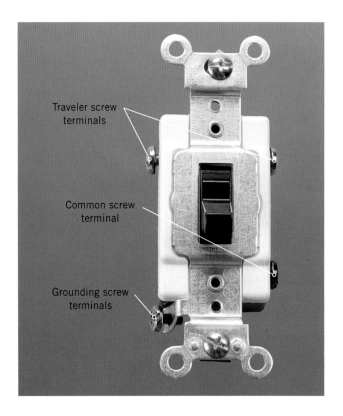

Traveler screw terminals

Common screw terminal

Grounding screw terminals

Typical Three-Way Switch Installation

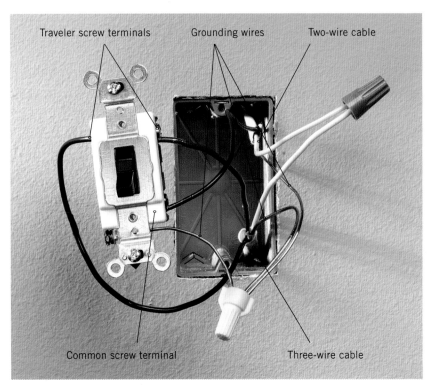

Traveler screw terminals   Grounding wires   Two-wire cable

Common screw terminal   Three-wire cable

Two cables enter the box: One cable has two wires, plus a bare copper grounding wire; the other cable has three wires, plus a ground. The black wire from the two-wire cable is connected to the dark common screw terminal. The red and black wires from the three-wire cable are connected to the traveler screw terminals. The white neutral wires are joined together with a wire connector, and the grounding wires are pigtailed to the switch grounding terminal.

How to Replace a Three-Way Wall Switch

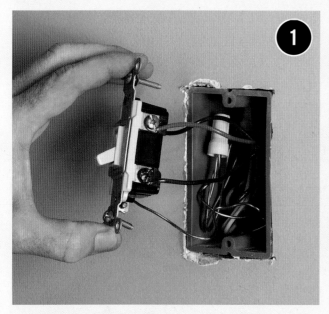

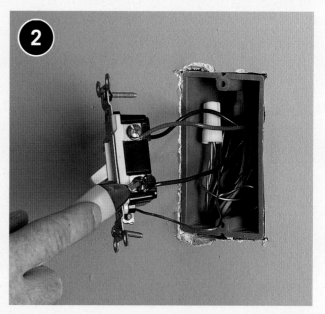

Turn off the power to the switch at the panel, and then remove the switch cover plate and mounting screws. Holding the mounting strap carefully, pull the switch from the box. Be careful not to touch the bare wires or screw terminals until they have been tested for power.

NOTE: If you are installing a new switch circuit, you must provide a neutral conductor at the switch.

Test for power by touching the probe of a noncontact voltage tester to each wire and screw terminal. Tester should not glow. If it does, there is still power entering the box. Return to the panel, and turn off the correct circuit, then test again for power.

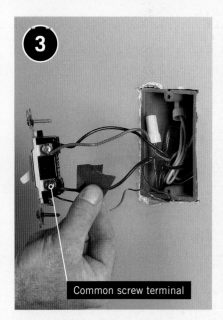

Common screw terminal

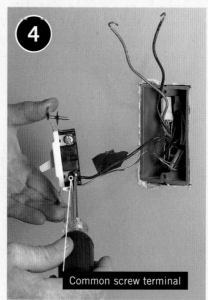

Common screw terminal

Locate the dark common screw terminal, and use masking tape to label the "common" wire attached to it. Disconnect wires and remove switch. Test the switch for continuity. If it tests faulty, buy a replacement. Inspect wires for nicks and scratches. If necessary, clip damaged wires and strip them.

Connect the common wire to the dark common screw terminal on the switch. On most three-way switches, the common screw terminal is black. Or it may be labeled with the word COMMON stamped on the back of the switch. Reconnect the grounding screw, and connect it to the circuit grounding wires with a pigtail.

Connect the remaining two circuit wires to the screw terminals. These wires are interchangeable and can be connected to either screw terminal. Carefully tuck the wires into the box. Remount the switch, and attach the cover plate. Turn on the power at the panel.

Four-Way Wall Switches

Four-way switches have four screw terminals and do not have ON-OFF markings. Four-way switches are always installed between a pair of three-way switches. This switch combination makes it possible to control a set of lights from three or more locations. Four-way switches are common in homes where large rooms contain multiple living areas, such as a kitchen opening into a dining room. Switch problems in a four-way installation can be caused by loose connections or worn parts in a four-way switch or in one of the three-way switches (facing page).

In a typical installation, there will be a pair of three-wire cables that enter the box for the four-way switch. With most switches, the black and red wires from one cable should be attached to the bottom or top pair of screw terminals, and the black and red wires from the other cable should be attached to the remaining pair of screw terminals. However, not all switches are configured the same way, and wiring configurations in the box may vary, so always study the wiring diagram that comes with the switch.

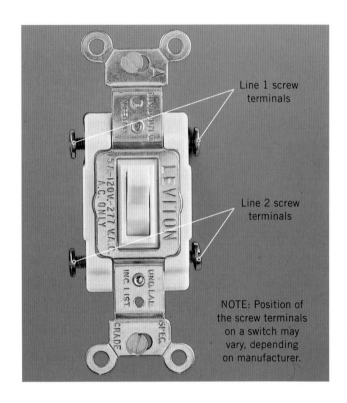

Line 1 screw terminals

Line 2 screw terminals

NOTE: Position of the screw terminals on a switch may vary, depending on manufacturer.

Common Four-Way Switch Installation

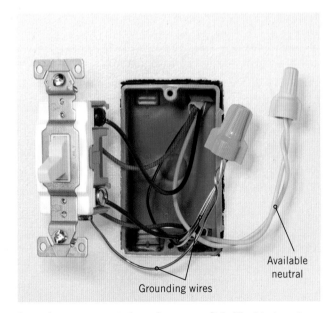

Available neutral

Grounding wires

Four wires are connected to a four-way switch. The black and red wires from one cable are attached to the top pair of screw terminals, while the black and red wires from the other cable are attached to the bottom screw terminals. In new switch wiring, the white wires are joined and bypass the switch but remain available for future use.

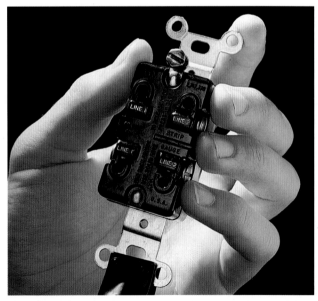

Switch variation: Some four-way switches have a wiring guide stamped on the back to help simplify installation. For the switch shown above, one pair of color-matched circuit wires will be connected to the screw terminals marked LINE 1, while the other pair of wires will be attached to the screw terminals marked LINE 2.

How to Replace a Four-Way Wall Switch

1

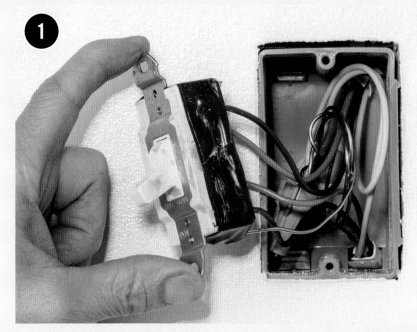

Turn off the power to the switch at the panel, and then remove the switch cover plate and mounting screws. Holding the mounting strap carefully, pull the switch from the box. Be careful not to touch any bare wires or screw terminals until they have been tested for power. Test for power by touching each wire and terminal with a noncontact voltage tester. The tester should not glow. If it does, there is still power entering the box. Return to the panel, and turn off the correct circuit.

2

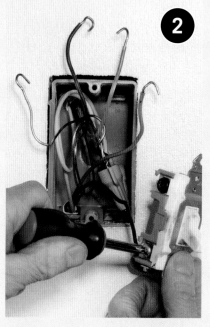

Disconnect the wires and inspect them for nicks and scratches. If necessary, clip damaged wires and strip them. Test the switch for continuity. Buy a replacement if the switch tests faulty.

3

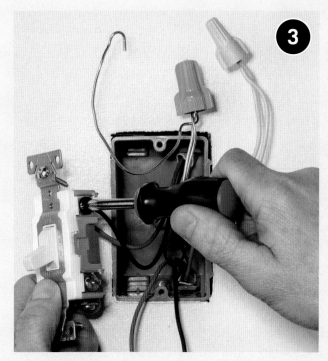

Connect two hot wires from one incoming cable to the top set of screw terminals.

4

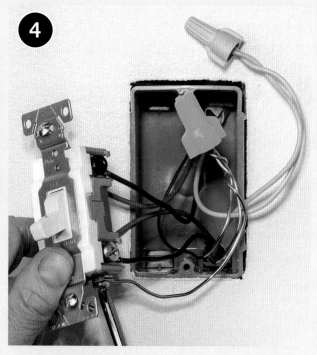

Attach remaining hot wires to the other set of screw terminals. Pigtail the grounding wires to the grounding screw. Carefully tuck the wires inside the switch box, and then remount the switch and cover plate. Turn on power at the panel.

Double Switches

A double switch has two switch levers in a single housing. It is used to control two light fixtures or appliances from the same switch box.

In most installations, both halves of the switch are powered by the same circuit. In these single-circuit installations, three wires are connected to the double switch. One wire, called the feed wire (which is hot), supplies power to both halves of the switch. The other wires, called the switch leg, carry power out to the individual light fixtures or appliances.

In rare installations, each half of the switch is powered by a separate circuit. In these separate-circuit installations, four wires are connected to the switch, and the metal connecting tab joining two of the screw terminals is removed (see photo below).

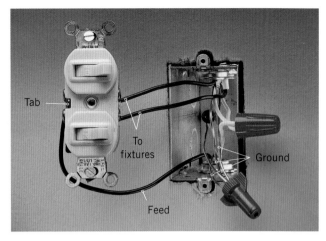

Single-circuit wiring: Three black wires are attached to the switch. The black feed wire bringing power into the box is connected to the side of the switch that has a connecting tab. The wires carrying power out to the light fixtures or appliances are connected to the side of the switch that does not have a connecting tab. The white neutral wires are connected together with a wire connector.

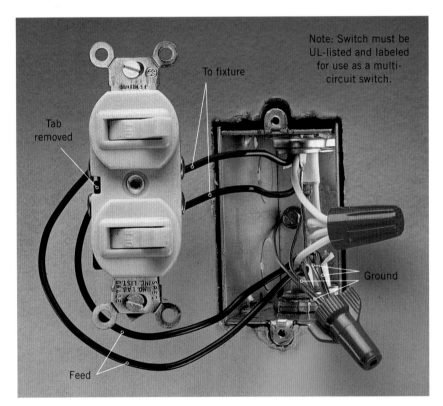

Separate-circuit wiring: Four black wires are attached to the switch. Feed wires from the power source are attached to the side of the switch that has a connecting tab, and the connecting tab is removed (photo, right). Wires carrying power from the switch to light fixtures or appliances are connected to the side of the switch that does not have a connecting tab. White neutral wires are connected together with a wire connector.

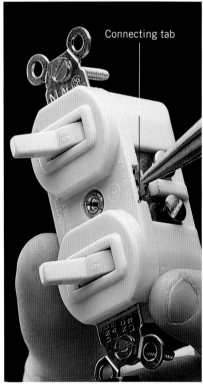

Remove the connecting tab on a double switch when wired in a separate-circuit installation. The tab can be removed with needlenose pliers or a screwdriver.

Pilot-Light Switches

A pilot-light switch has a built-in bulb that glows when power flows through the switch to a light fixture or appliance. Pilot-light switches often are installed for convenience if a light fixture or appliance cannot be seen from the switch location. Basement lights, garage lights, and attic exhaust fans frequently are controlled by pilot-light switches.

A pilot-light switch requires a neutral wire connection. A switch box that contains a single two-wire cable has only hot wires and cannot be fitted with a pilot-light switch.

Switch/Receptacles

A switch/receptacle combines a grounded receptacle with a single-pole wall switch. In a room that does not have enough wall receptacles, electrical service can be improved by replacing a single-pole switch with a switch/receptacle.

A switch/receptacle requires a neutral wire connection. A switch box that contains a single two-wire cable has only hot wires and cannot be fitted with a switch/receptacle.

A switch/receptacle can be installed in one of two ways. In the most common installations, the receptacle is hot even when the switch is off (photo, right).

In rare installations, a switch/receptacle is wired so the receptacle is hot only when the switch is on. In this installation, the hot wires are reversed, so that the feed wire is attached to the brass screw terminal on the side of the switch that does not have a connecting tab.

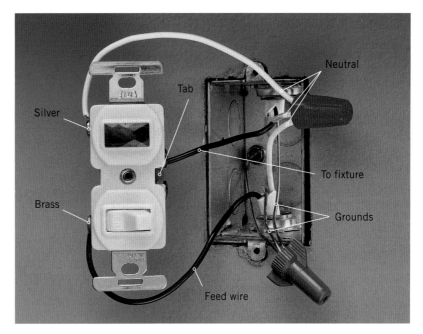

Pilot-light switch wiring: Three wires are connected to the switch. One black wire is the feed wire that brings power into the box. It is connected to the brass (gold) screw terminal on the side of the switch that does not have a connecting tab. The white neutral wires are pigtailed to the silver screw terminal. The black wire carrying power out to a light fixture or appliance is connected to the screw terminal on the side of the switch that has a connecting tab.

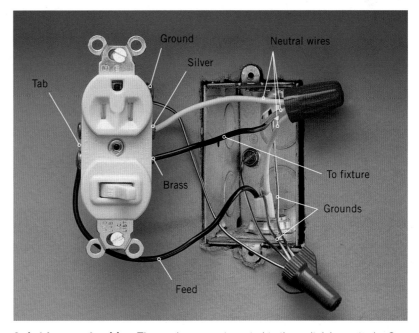

Switch/receptacle wiring: Three wires are connected to the switch/receptacle. One of the hot wires is the feed wire that brings power into the box. It is connected to the side of the switch that has a connecting tab. The other hot wire carries power out to the light fixture or appliance. It is connected to the brass screw terminal on the side that does not have a connecting tab. The white neutral wire is pigtailed to the silver screw terminal. The grounding wires must be pigtailed to the green grounding screw on the switch/receptacle and to the grounded metal box.

Specialty Switches

Your house may have several types of specialty switches. Dimmer switches are used frequently to control light intensity in dining and recreation areas. Timer switches and time-delay switches (below) are used to control light fixtures and exhaust fans automatically. Electronic switches provide added convenience and home security, and they are easy to install. Electronic switches are durable, and they rarely need replacement.

Most specialty switches have preattached wire leads instead of screw terminals and are connected to circuit wires with wire connectors. Some motor-driven timer switches require a neutral wire connection and cannot be installed in switch boxes that have only one cable with two hot wires. It is precisely due to the rise in popularity of "smart" switches that the NEC Code was changed in 2014 to require an available neutral wire in newly installed switch boxes.

If a specialty switch is not operating correctly, you may be able to test it with a continuity tester. Timer switches and time-delay switches can be tested for continuity, but dimmer switches cannot be tested. With electronic switches, the manual switch can be tested for continuity, but the automatic features cannot be tested.

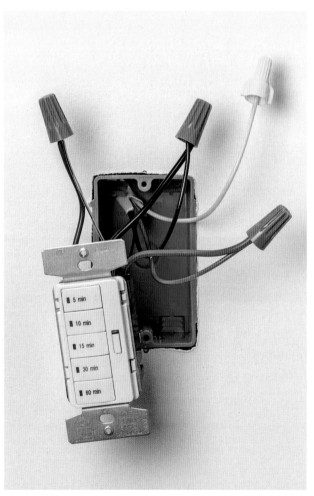

Countdown timer switch. This rocker-type switch gives you the option to easily program the switch to shut off after a specified time: from 5 to 60 minutes. Garage lights or basement lights are good applications: anywhere you want the light to stay on long enough to allow you to exit, but not to stay on indefinitely. These switches often are used to control exhaust fans.

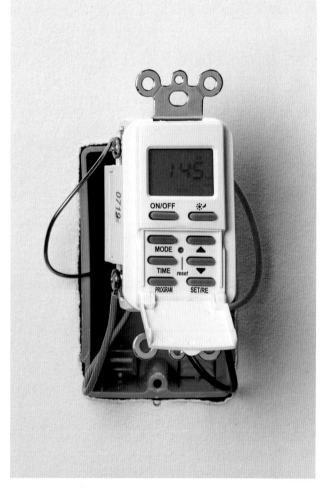

Programmable timer switch. A digital or dial-type timer allows you to program the switch to turn on for specific time periods at designated times of day within a 24-hour cycle. Security lights, space heaters, towel warmers, and radiant floors are typical applications.

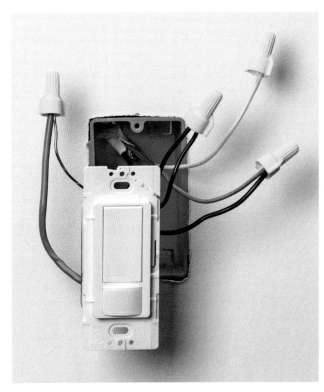

Occupancy sensor. Many smart switches incorporate a motion detector that will switch the lights on if they sense movement in the room and will also shut them off when no movement is detected for a period of time. The model shown above also has a dimmer function for further energy savings.

Spring-wound timer switch. A relatively simple device, this timer switch functions exactly like a kitchen timer, employing a hand-turned dial to and spring mechanism to shut the switch off in increments up to 15 minutes.

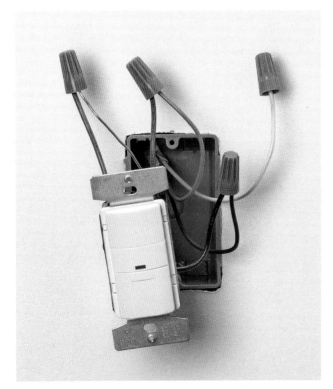

Daylight sensor switch. This switch automatically turns on when light levels drop below a proscribed level. It can also be programmed as an occupancy sensor to shut off when the room is vacant and turn on when the room is entered.

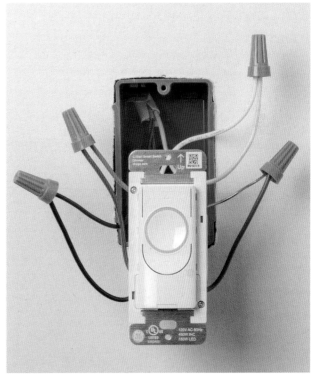

Smart switches let you control lights from a smart phone or other compatible device, such as a smart speaker, without the need for a central hub (a common requirement with early smart home systems).

Ground-fault (GFCI) & Arc-fault (AFCI) Protection

Ground-Fault Location Requirements

1. Kitchen receptacles. Install ground-fault circuit interrupt (GFCI) protection on all 120-volt and 240-volt receptacles that serve kitchen countertops. This does not include receptacles under the kitchen sink receptacles located on kitchen walls that do not serve the countertop and receptacles that are not within six feet of a sink.

2. Kitchen. Install ground-fault circuit interrupt (GFCI) protection on the outlets that supply dishwashing machines.

3. Bathroom receptacles. Install ground-fault circuit interrupt (GFCI) protection on all 120-volt and 240-volt receptacles located in bathrooms. This applies to all receptacles regardless of where they are located in the bathroom and includes receptacles located at countertops, inside cabinets, and along bathroom walls. This also applies to bathtubs and shower stalls that are not located in a bathroom. Install ground-fault circuit interrupt (GFCI) protection on all circuits serving electrically heated floors in bathrooms, kitchens, and around whirlpool tubs, spas, and hot tubs.

4. Garage and Accessory Building receptacles. Install ground-fault circuit interrupt (GFCI) protection on all 120-volt and 240-volt receptacles located in garages and grade-level areas of unfinished accessory buildings.

5. Exterior receptacles. Install ground-fault circuit interrupt (GFCI) protection on all 120-volt and 240-volt receptacles located outdoors. This does not apply to receptacles that are dedicated for deicing equipment and are located under the eaves. This applies to holiday lighting receptacles located under the eaves.

6. Basement receptacles. Install ground-fault circuit interrupt (GFCI) protection on all 120-volt and 240-volt receptacles located in unfinished

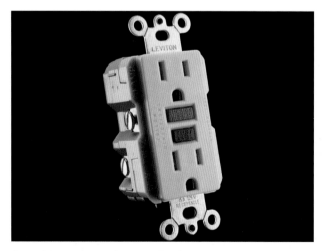

Ground-fault receptacles and circuit breakers detect unwanted current running between an energized wire and a grounded wire.

A combination ARC-fault circuit breaker detects sparking (arcing) faults along damaged energized wires and detects these faults between wires. A branch ARC-fault circuit breaker only detects arcing faults between wires.

Receptacles for whirlpool tubs must be GFCI protected.

basements. An unfinished basement is not intended as habitable space and is limited to storage and work space.

7. Crawl space receptacles. Install ground-fault circuit interrupt (GFCI) protection on all 120-volt and 240-volt receptacles located in crawl spaces. Receptacles in crawl spaces are not required unless equipment requiring service is located there.

8. Sink receptacles. Install ground-fault circuit interrupt (GFCI) protection on all 120-volt and 240-volt receptacles that are located within six feet of the outside edge of a sink. This includes wall, floor, and countertop receptacles.

9. Boathouse receptacles. Install ground-fault circuit interrupt (GFCI) protection on all 120-volt and 240-volt receptacles located in boathouses.

10. Spas, tubs, and other circuits requiring ground-fault protection. Install ground-fault circuit interrupt (GFCI) protection on all circuits serving spa tubs, whirlpool tubs, hot tubs, and similar equipment. Refer to the general codes for more information about circuits serving these components.

11. Install GFCI circuit breakers and receptacles so that they are readily accessible.

Arc-Fault Location Requirements

1. Install a combination type or an outlet (receptacle) type arc-fault circuit interrupter (AFCI) on all

15- and 20-amp, 120-volt branch circuits serving sleeping, family, dining, living, sun, and recreation rooms, kitchens, laundry areas, and parlors, libraries, dens, hallways, closets, and similar rooms and areas. This means that 15-and 20-amp, 120-volt branch circuits serving most interior spaces in a home are required to have AFCI protection. Note that garages, basements, utility and mechanical rooms, and exterior branch circuits are not included in this list, although local building officials may include these areas by interpretation.

2. You may provide AFCI protection for the entire branch circuit by installing a combination-type AFCI circuit breaker in the electrical panel where the branch circuit originates.

3. You may provide AFCI protection to a branch circuit using several different combinations of branch-circuit type AFCI circuit breakers and branch-circuit type AFCI receptacles. Refer to general codes or your local building inspector for details about these alternate methods.

4. Provide AFCI for branch circuits that are modified, replaced, or extended. You may use either of the following methods: (a) install a combination-type AFCI circuit breaker in the electrical panel where the branch circuit originates, or (b) install a branch-circuit type AFCI receptacle at the first receptacle in the existing branch circuit.

5. Install AFCI circuit breakers and receptacles so that they are readily accessible.

Junction Boxes, Device Boxes & Enclosures

All electrical boxes are available in different depths. A box must be deep enough so a switch or receptacle can be removed or installed easily without crimping and damaging the circuit wires. Replace an undersized box with a larger box using the Electrical Box Fill Chart (see page 124) as a guide. **The NEC also says that all electrical boxes must remain accessible. Never cover an electrical box with drywall, paneling, or wall coverings.**

Nonmetallic Box Installation

1. Use nonmetallic boxes only with NM type cable or with nonmetallic conduit or tubing. You may use nonmetallic boxes with metallic conduit or tubing if you maintain the electrical continuity of the metallic conduit or tubing by installing a bonding jumper through the box. In many situations it is easier to use a metallic box with metallic conduit or tubing.

2. Extend NM cable sheathing at least ¼ inch into a nonmetallic box knockout opening.

3. Secure NM cable, conduit, and tubing to each box. You may secure NM cable with cable clamps inside the box or with compression tabs provided where the cable enters the box. You do not need to secure NM cable to a standard single-gang box (2¼ by 4 inches) mounted in a wall or ceiling if you fasten the cable not more than eight inches from the box and if the sheathing enters the box at least ¼ inch. Measure the eight inches along the length of the sheathing, not from the outside of the box.

Light Fixture Box Installation

1. Use boxes designed for mounting light fixtures if a light fixture is to be mounted to the box. These boxes are usually four-inch round or octagonal.

2. You may use other boxes to mount light fixtures on walls if the fixture weighs less than 6 pounds and if the fixture is secured to the box using at least #6 screws.

3. Support light fixtures weighing at least 50 pounds independently from the light fixture box. You may use the light fixture box to support light fixtures weighing less than 50 pounds. Note that ceiling fans are not light fixtures.

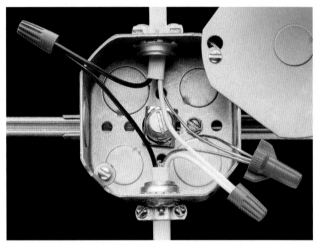

Box shape is directly related to function, as electrical fixtures are created to fit on boxes of a particular shape. Octagonal and round boxes generally are designed for ceiling mounting, while square and rectangular boxes are sized for single-pole, duplex, and other standard switch and receptacle sizes.

Do not support heavy light fixtures using only the light fixture electrical box. The eye hook supporting this chandelier is driven into the same ceiling joist to which the electrical box is mounted.

Box Contents Limitations

1. Limit the number of wires, devices (such as switches and receptacles), and fittings in a box. This limitation is primarily based on the heat generated by the wires and devices in the box. The actual size of the box relative to its contents is a secondary consideration.

2. Use the cubic inch volume printed on the box or provided in the box manufacturer's instructions to determine box volume. Do not attempt to measure the box volume. Do not estimate box volume from the volume of similar size boxes. You will probably not get the same volume as provided by the manufacturer.

3. Use table "Wire Volume Unit" to determine the volume units required by wires, devices, and fittings in a box.

Box Installation Tolerances

1. Install boxes in non-combustible material, such as masonry, so that the front edge is not more than ¼ inch from the finished surface.

2. Install boxes in walls and ceilings made of wood or other combustible material so that the front edge is flush with the finished surface or projects from the finished surface.

3. Cut openings for boxes in drywall and plaster so that the opening is not more than ⅛ inch from the perimeter of the box.

Boxes must be installed so the front edges are flush with the finished wall surface, and the gap between the box and the wall covering is not more than ⅛".

WIRE VOLUME UNIT

WIRE SIZE (AWG)	WIRE VOLUME
14	2.00 in.3
12	2.25 in.3
10	2.50 in.3
8	3.00 in.3
6	5.00 in.3

VOLUME UNITS

Calculate the volume units required by wires, devices, and fittings based on the following definitions:

Volume units for current-carrying wires. Allow one volume unit for each individual hot (ungrounded) and neutral (grounded) wire in the box. Use Table 47 to determine the volume units of common wire sizes. Example: two pieces of #14/2 NM are in a box. Each piece of this cable contains one hot (ungrounded) and one neutral (grounded) wire and one grounding wire. From table "Wire Volume Unit", each #14 wire uses 2.00 cubic inches in the box. The total volume units required by the hot (ungrounded) and neutral (grounded) wires is eight cubic inches.

Volume units for devices. Allow two volume units for each device (switch or receptacle) in the box. Base the volume units on the largest hot (ungrounded) or neutral

(grounded) wire in the box. Example: NM cable size #14 and #12 are in a box. From Table 47, #14 wire uses 2.00 cubic inches and #12 wire uses 2.25 cubic inches. Allow 4.5 cubic inches volume units (2 × 2.25 cubic inches) for each switch or receptacle in the box based on the volume of the larger #12 NM cable.

Volume units for grounding wires. Allow one volume unit for all grounding wires in the box. Base the volume unit on the largest hot (ungrounded) or neutral (grounded) wire in the box.

Volume units for clamps. Allow one volume unit for all internal cable clamps in the box, if any. Base the volume unit on the largest hot (ungrounded) or neutral (grounded) wire in the box.

Volume units for fittings. Allow one volume unit for all fittings in the box, if any. Base the volume unit on the largest hot (ungrounded) or neutral (grounded) wire in the box.

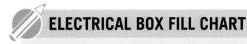

ELECTRICAL BOX FILL CHART

BOX SIZE AND SHAPE	MAXIMUM NUMBER OF CONDUCTORS PERMITTED (SEE NOTES BELOW)			
WIRE SIZE	8 AWG	10 AWG	12 AWG	14 AWG
JUNCTION BOXES				
4 × 1¼" R or O	5	5	5	6
4 × 1½" R or O	5	6	6	7
4 × 2⅛" R or O	7	8	9	10
4 × 1¼" S	6	7	8	9
4 × 1½" S	7	8	9	10
4 × 2⅛" S	10	12	13	15
4¹¹⁄₁₆ × 1¼" S	8	10	11	12
4¹¹⁄₁₆ × 1½" S	9	11	13	14
4¹¹⁄₁₆ × 2⅛" S	14	16	18	21
DEVICE BOXES				
3 × 2 × 1½"	2	3	3	3
3 × 2 × 2"	3	4	4	5
3 × 2 × 2¼"	3	4	4	5
3 × 2 × 2½"	4	5	5	6
3 × 2 × 2¾"	4	5	6	7
3 × 2 × 3½"	6	7	8	9
4 × 2⅛ × 1½"	3	4	4	5
4 × 2⅛ × 1⅞"	4	5	5	6
4 × 2⅛ × 2⅛"	4	5	6	7

NOTES:
- R = Round; O = Octagonal; S = Square or rectangular
- Each hot or neutral wire entering the box is counted as one conductor.
- Grounding wires are counted as one conductor in total—do not count each one individually.
- Raceway fittings and external cable clamps do not count. Internal cable connectors and straps count as either half or one conductor, depending on type.
- Devices (switches and receptacles mainly) each count as two conductors.
- When calculating total conductors, any nonwire components should be assigned the gauge of the largest wire in the box.
- For wire gauges not shown here, contact your local electrical inspections office.

Box Support in Walls, Ceilings & Floors

1. Provide support for boxes that rigidly and securely fasten them in place. You may use nails or screws to support these boxes.

2. Protect screws inside boxes so that the threads will not damage the wires.

3. Wood braces used to support boxes must be at least one by two inches.

4. Use "cut-in" or "old work" retrofit boxes only if they have approved clamps or anchors that are identified for the location where they are installed.

Damp Locations

1. Install a receptacle box cover that is weatherproof when the cover is closed and a plug is not inserted into a receptacle located in a damp location. This applies to 120-volt, 240-volt, 15-amp, and 20-amp receptacles. A damp area is protected from direct contact with water. Refer to the definition of damp location. You may use a receptacle cover suitable for wet locations in a damp location.

2. Install a watertight seal between a flush-mounted receptacle and its faceplate. This will require a gasket or sealant between the finished surface (such as stucco, brick, or siding) and the faceplate.

Wet Locations

1. Install a receptacle box cover that is weatherproof when the cover is closed on any receptacle located in a wet location. This applies to 120-volt, 240-volt, 15-amp, and 20-amp receptacles in any indoor or outdoor wet location. This applies regardless of whether or not a plug is inserted into the receptacle.

2. Install a watertight seal between a flush-mounted receptacle and its faceplate. This will require a gasket or sealant between the finished surface (such as stucco, brick, or siding) and the faceplate.

Resources

National Renewable Energy Laboratory
Renewable energy research
303 384 6565
www.NREL.gov

Black+Decker
Portable power tools and more
www.blackanddecker.com

Broan-NuTone, LLC
Vent fans
800 558 1711
www.broan.com

Generac Power Systems
Standby generators and switches
888 436 3722
www.generac.com

Honda Power Equipment/American Honda Motor Company, Inc.
Standby generators
770 497 6400
www.hondapowerequipment.com

Kohler
Standby generators
800 544 2444
www.kohlergenerators.com

Pass & Seymour Legrand
Home automation products
877 295 3472
www.passandseymour.com

Westinghouse
Ceiling fans, decorative lighting, solar outdoor
   lighting & other lighting fixtures and bulbs
866 442 7873
Purchase here: www.westinghouse.com

Photo Credits

p. 16 and 17 (top left) Shutterstock
p. 53 photo © Mike Clarke / www.istock.com
p. 74 photo © Jeff Chevrier / www.istock.com
p. 75 photos (top right & lower) courtesy of Generac Power
   Systems, Inc.

p. 86 photo courtesy of Cabin Fever, featuring McMaster Carr
   vapor-tight light fixtures
p. 111 (bottom left, bottom right), 114 (bottom left), 115, 118, 119
   Robert B. Bartee

Conversions

METRIC EQUIVALENT

	1/64	1/32	1/25	1/16	1/8	1/4	3/8	2/5	1/2	5/8	3/4	7/8	1	2	3	4	5	6	7	8	9	10	11	12	36	39.4
Inches (in.)	1/64	1/32	1/25	1/16	1/8	1/4	3/8	2/5	1/2	5/8	3/4	7/8	1	2	3	4	5	6	7	8	9	10	11	12	36	39.4
Feet (ft.)																								1	3	3 1/12
Yards (yd.)																									1	1 1/12
Millimeters (mm)	0.40	0.79	1	1.59	3.18	6.35	9.53	10	12.7	15.9	19.1	22.2	25.4	50.8	76.2	101.6	127	152	178	203	229	254	279	305	914	1,000
Centimeters (cm)							0.95	1	1.27	1.59	1.91	2.22	2.54	5.08	7.62	10.16	12.7	15.2	17.8	20.3	22.9	25.4	27.9	30.5	91.4	100
Meters (m)																								.30	.91	1.00

CONVERTING MEASUREMENTS

TO CONVERT:	TO:	MULTIPLY BY:
Inches	Millimeters	25.4
Inches	Centimeters	2.54
Feet	Meters	0.305
Yards	Meters	0.914
Miles	Kilometers	1.609
Square inches	Square centimeters	6.45
Square feet	Square meters	0.093
Square yards	Square meters	0.836
Cubic inches	Cubic centimeters	16.4
Cubic feet	Cubic meters	0.0283
Cubic yards	Cubic meters	0.765
Pints (U.S.)	Liters	0.473 (Imp. 0.568)
Quarts (U.S.)	Liters	0.946 (Imp. 1.136)
Gallons (U.S.)	Liters	3.785 (Imp. 4.546)
Ounces	Grams	28.4
Pounds	Kilograms	0.454
Tons	Metric tons	0.907

TO CONVERT:	TO:	MULTIPLY BY:
Millimeters	Inches	0.039
Centimeters	Inches	0.394
Meters	Feet	3.28
Meters	Yards	1.09
Kilometers	Miles	0.621
Square centimeters	Square inches	0.155
Square meters	Square feet	10.8
Square meters	Square yards	1.2
Cubic centimeters	Cubic inches	0.061
Cubic meters	Cubic feet	35.3
Cubic meters	Cubic yards	1.31
Liters	Pints (U.S.)	2.114 (Imp. 1.76)
Liters	Quarts (U.S.)	1.057 (Imp. 0.88)
Liters	Gallons (U.S.)	0.264 (Imp. 0.22)
Grams	Ounces	0.035
Kilograms	Pounds	2.2
Metric tons	Tons	1.1

CONVERTING TEMPERATURES

Convert degrees Fahrenheit (F) to degrees Celsius (C) by following this simple formula: Subtract 32 from the Fahrenheit temperature reading. Then mulitply that number by $5/9$. For example, 77°F - 32 = 45. 45 × $5/9$ = 25°C.

To convert degrees Celsius to degrees Fahrenheit, multiply the Celsius temperature reading by $9/5$, then add 32. For example, 25°C × $9/5$ = 45. 45 + 32 = 77°F.

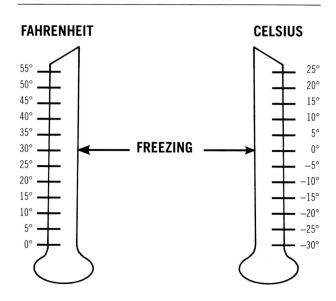

Index